ESSENTIALS
OF SOIL STUDY

With special reference to tropical areas

Dychweler y llyfr hwn erbyn y dyddiad isod, neu pan elwir amdano gan y Llyfrgellydd. Rhoddir dirwy o ~~2½p~~ y dydd ar unrhyw ddarllenydd a fetho ddychwelyd y llyfr erbyn y dyddiad hwn.

This book must be returned on or before the date shown below, or when required by the Librarian. A fine of ~~2½p~~ per day will be imposed on any borrower failing to return the book by this date.

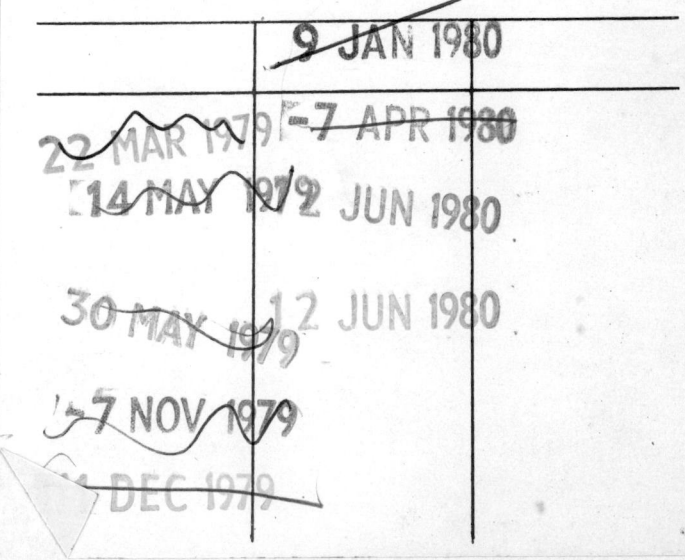

ESSENTIALS OF SOIL STUDY

With special reference to tropical areas

A. Faniran
Reader in Geography at the University of Ibadan

O. Areola
Lecturer in Geography at the University of Ibadan

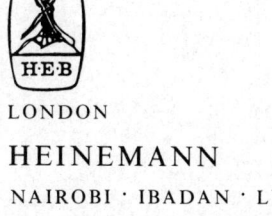

LONDON
HEINEMANN
NAIROBI · IBADAN · LUSAKA

Heinemann Educational Books Ltd
48 Charles Street, London W1X 8AH
P.M.B. 5205 Ibadan · P.O. Box 45314 Nairobi · P.O. Box 3966 Lusaka

EDINBURGH MELBOURNE TORONTO AUCKLAND
SINGAPORE HONG KONG KUALA LUMPUR
NEW DELHI KINGSTON

ISBN 0 435 95311 7 (paper)

© A. Faniran and O. Areola 1978
First published 1978

Text set in 10 pt Times Roman
Printed by Butler & Tanner Ltd, Frome and London

CONTENTS

	page
List of Tables	vii
List of Figures	viii
Preface	ix

Chapter 1 INTRODUCTION   1
Living Soil   1
Pedology and Geography   1
History of the Development of Pedology and Concepts of Soil   3
Definition and Study of Soil   7

Chapter 2 SOIL CONSTITUENTS AND PROPERTIES   9
Soil Constituents   9
Soil Properties   17
The Soil Profile   29

Chapter 3 SOIL FORMATION I: ROCK WEATHERING   33
The Process of Rock Weathering   33
The Weathering Crust   34
Types of Weathering   34
Factors Influencing Rock Weathering   42
Energy and Rock Weathering   45
The Product of Rock Weathering   45

Chapter 4 SOIL FORMATION II: SOIL-FORMING FACTORS AND PROCESSES   47
The Factors of Soil Formation   48
The Processes of Soil Formation   65

Chapter 5 SOIL SURVEY TECHNIQUES AND PROCEDURES   78
Soil Survey Organizations   78
Stages in a Soil Survey Project   80
Soil Survey Interpretation   86
Soil in a Multidisciplinary Integrated Land Resources Survey   89
Soil Survey with Special Reference to Africa   91

Chapter 6 METHODS OF STUDYING THE SOIL PROFILE IN THE FIELD 97
Soil Survey Equipment 97
Stages in the Study of a Soil Profile in the Field 98
Soil Monolith 121

Chapter 7 SOIL CLASSIFICATION 124
The Basis of Soil Classification 125
The Soil Map of Africa 144
Soil Classification for Specific Purposes 145
A Suggested System for Soil Classification 149

Chapter 8 WORLD SOILS 156
The Soils of the Humid Tropics 157
The Soils of the Humid Temperate Regions 172
The Soils of the Semi-Arid and Arid Regions 179
The Soils of the Cold Lands 184
Relict Soils 186
Regosols, Lithosols, and Raw Mineral Soils 191

Chapter 9 SOIL–PLANT RELATIONS 192
Factors of Soil Fertility 192
Problems of Soil Erosion and Pollution 204

Chapter 10 SOIL EROSION AND CONSERVATION 206
The Nature of Soil Erosion 207
The Factors of Soil Erosion 208
Agents of Soil Erosion 218
Soil Conservation Efforts 228

Appendix LABORATORY TECHNIQUES OF SOIL ANALYSIS 237
Laboratory Equipment and Organization 238
Methods of Soil Analysis 238

References 267

Index 273

LIST OF TABLES

		page
2.1	Composition of soil air compared to the atmosphere	15
3.1	Expansion factors of some common rock minerals	36
4.1	Hydrological soil series	53
4.2	The processes and reactions of soil formation	70
5.1	Comparison of textural classes, western Nigeria: USDA	94
5.2	The USDA land capability classification system	95
6.1	Soil survey proforma:	
	(a) soil site characteristics	99
	(b) soil profile characteristics	100
6.2	Measurement of slopes: table of equivalents	102
6.3	Soil textural classes: definition and identification	109
6.4	Stone class determination	113
6.5	Names used for coarse fragments in soils	113
6.6	Determination of percentage free carbonate content	122
7.1	World soils according to G. W. Robinson (1947)	127
7.2	Classification of soils: the great soil group system	129
7.3	A comparison of botanical and soil classification systems	130
7.4	Soil classification according to Stephens (1962)	132
7.5	The soil classification system of Stace, Hubble, and others (1968)	134
7.6	(a) The new American soil classification system: summary of the global soil orders	135
	(b) Definitions of some diagnostic horizons of the new American soil classification system	138
7.7	The French system of soil classification	140
7.8	Kubiena's (1953) system of soil classification	137
7.9	Avery's (1956) classification of British soils	142
7.10	Legend of 1:5,000,000 soil map of Africa	146
7.11	Simple soil-forming processes	150
7.12	Relationship between soil groups used in this book and the other popular systems of soil classification	152
8.1	Chemical analysis of duricrusts and other weathered products from the Sydney area	162
8.2	Polished section modal analyses of laterites in fine-grained sandstone, Sydney district	163
8.3	Depth of weathering in Nigeria: some statistical measures	167
8.4	Deep weathering in Nigeria: profile analysis	168
10.1	Effect of slope and cultural practices on soil and water loss (Ibadan IITA, 1972)	214
10.2	Relationship between strip width and soil textural class	235
A.1	Table of settling times for separation of fractions	250

LIST OF FIGURES

		page
1.1	The relations of pedology	2
2.1	Volume composition of a silty-loam surface soil when in ideal condition for plant growth	9
2.2	Systems of texture grades	19
2.3	Types of soil structure	20
2.4	Schematic representation of the lattice structure of the 1 : 1 kaolin and 2 : 1 montmorillonite clays	25
2.5	A clay crystal with excess negative charges and adsorbed cations	25
2.6	The soil profile and horizon nomenclature	30
2.7	The movement of bases in a forest soil	31
3.1	Types of weathering	35
3.2	Weathering sequence of common rock types	43
3.3	Topography, groundwater, and weathering zones	44
4.1	Soil water balance	51
4.2	Precipitation and soil formation	52
4.3	Relationship between slope steepness and depth of eluvial horizons	57
4.4	Changes in solum thickness with slope angle (Mudgee area, Australia)	58
4.5	Effect of slope aspect	59
4.6	Soil catena in a humid region in Nigeria's Basement Complex	60
4.7	Mountains and soil zonation	62
4.8	Time factor in soil formation	64
4.9	Topographic dissection and soils	65
4.10	The process of leaching	62
4.11	Climate and the process of humification	68
4.12	Pedogenic zones	72
4.13	Processes of soil formation	73
4.14	Climate–soil–vegetation relationships	74
5.1	A triangular grid	84
6.1	A simple hillslope model showing morphological divisions	103
6.2	Soil textural classes	111
8.1	New American comprehensive soil classification system	156
8.2	The humid tropics	159
8.3	Examples of podzol profile	175
8.4	The major occurrences of duricrust	188
8.5	World soils—a simplified picture	190
9.1	The nitrogen cycle	197
9.2	The phosphorus cycle	199
9.3	The sulphur cycle	202
10.1	Factors of soil erosion by water	210

PREFACE

In the past there has been so much confusion about the nature of the soil that textbooks on soils having the same title have often differed widely in the subject matter treated. Happily, pedology or soil science has developed beyond that stage. Soil is such a basic element of land that it has remained a subject of considerable interest to many disciplines. Unfortunately each discipline has tended to emphasize only those aspects of the study of the soil most relevant to it. Thus, there are very few textbooks which look at the soil from a broad viewpoint or give a balanced treatment of its various aspects. The geographer, by the very nature of his discipline, has to view the soil from such a broad, all-embracing perspective: as a major element of land which determines some of its most essential qualities; as the product of the interaction and an essential link between the lithosphere, the atmosphere, the hydrosphere, and the biosphere; and as a resource and a component of the environment system or ecosystem. This is what this book aims to do. Although it has been written specifically for soil geographers at university level, where the subject is being studied seriously, we hope it will be just as useful to scientists and students in other disciplines such as forestry, geology, agronomy, and archaeology, to name a few.

In writing this book, we have been guided by certain basic principles. First, we have isolated for discussion those topics that we consider relevant for the effective study of the soil, no matter by whom. Secondly, we have tried to correct the deficiency in most of the existing textbooks, which is a paucity of examples from the tropical world in general and Africa in particular. African students should therefore feel more at home with this book. Thirdly, we have tried to write as simply and as clearly as possible, using the best examples known to us. At the same time, no attempt is made to oversimplify difficult areas. Rather, the problematic nature of many aspects of soil study is brought out. Finally, we have emphasized that soils, like other natural phenomena, result from the interaction of many factors and processes all of which have to be analysed and assessed both in the field and in the laboratory. The scientific study of soils can only be accomplished by well-organized field surveys complemented by detailed study of air photographs, and the analysis of samples in well-equipped laboratories.

The book consists of ten chapters. Chapters 1 to 4 describe the soil body, its properties, and formation. Chapters 5 and 6 deal with the study of soil both by itself and with other related resources during integrated survey programmes. Chapters 7 and 8 consider soil types and their global distribution. Chapter 9 is on soil–plant relations and Chapter 10 on soil erosion and conservation. The appendix complements the chapters on soil study by describing the apparatus and procedures of the laboratory study of soils.

We are grateful to the Department of Geography, University of Ibadan, for providing typing and cartographic facilities, and to our wives and children for tolerating our long periods of absence from home.

ADETOYE FANIRAN
OLUSEGUN AREOLA

Chapter 1

INTRODUCTION

Living Soil

Soil is the major element of land, the solid part of the earth's surface. It is derived from rocks by physical, chemical, and biological processes. Although the soil originates from rocks it is essentially different from them in certain respects. Unlike rocks, the soil is made up largely of new, secondary minerals. These secondary minerals are continually being formed from the underlying rocks through the processes of weathering and pedogenesis, while at the same time the top part of the soil is continually being removed by denudation. The soil, therefore, rather than being a static entity, is always in a state of flux and change. Also, the soil has a relatively high organic matter content in the form of organic compounds, plant roots, and soil organisms. These organic constituents make the soil come alive. *The living soil* is the medium for plant growth on which man depends for his food supply. Furthermore, soils, unlike rocks, are regularly distributed over the earth's surface, showing distinct spatial relationships with other factors of the environment. The soil thus constitutes a real entity in the natural environment whose character, development, and distribution in space are governed by the general laws of nature. This view of the soil is the basis of the scientific discipline known as pedology.

Pedology and Geography

Pedology is commonly defined as the branch of natural or earth science which studies the origin, properties, and distribution of soils. Since soil fertility is intimately related to these pedological attributes, pedology also studies the best uses of soil and is involved in the all-important problem of natural resource inventory and evaluation, especially as it relates to soils. As such, pedology encompasses a wide, if not the entire, field of soil science. Indeed, the pedologist in most cases relies on, as well as feeds back information to, the cognate fields such as soil chemistry, soil physics, soil microbiology, etc. To this extent pedology is like geography, which not only synthesizes the various specialist areas of knowledge about the earth as the home of man, but also contributes knowledge to the specialist fields in the natural, social, and even fundamental (geometrical) sciences (Fig. 1.1).

Pedology also seems to have a special appeal to the geographer, especially the physical geographer or earth scientist, for other reasons. Soil is

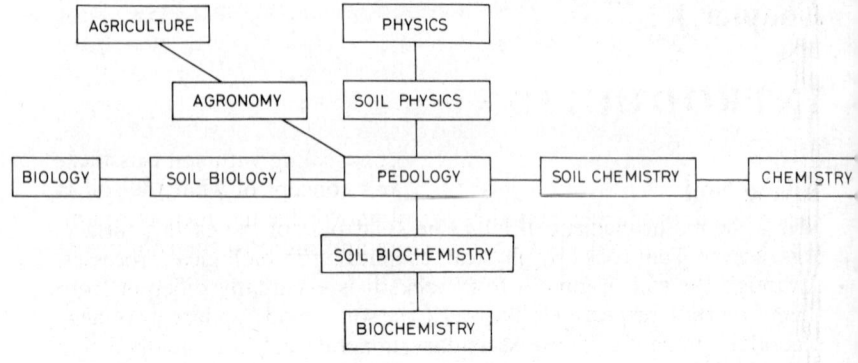

Fig. 1.1 The relations of pedology

a part of the physical environment which the physical geographer studies. It is a typically geographic topic, resulting from and reflecting the interaction of the elements of the earth's surface, including the atmospheric, the lithospheric, the hydrospheric, and the biospheric elements. This means that soil can only be understood when it is considered along with the other sciences which study the earth's surface, namely, meteorology, climatology, geomorphology, biogeography, etc., all of which also lie within the orbit of physical geography. It is therefore not surprising that the best pedologists, both past and present, include geographers or people with adequate backgrounds in the basic geographical concepts and methods.

The methodology both of pedology and of geography is also similar. First, the approach of the pedologist in studying soil sites and soil profiles, soil-forming factors and processes, and soil distribution is akin to that of the geographer studying points, patterns, and processes in relation to earth-surface phenomena. The unifying concepts of pedology and geography are also similar in some respects. Both operate in space and both study among other things the delicate interplay between natural forces and factors on the one hand and the effects of activities of organisms including man on the other. Finally, the history of both disciplines has followed a similar course. The development of geomorphology, plant geography, and pedology alike is marked by a long initial period of uncertainty, false concepts, and groping in the dark, followed by a much shorter period during which the basic principles and methods of the subjects as we know and practise them today were formulated. The case of pedology is considered briefly next; parallels are drawn from either of these other natural sciences.

History of the Development of Pedology and Concepts of Soil

The development of *pedology* as a real scientific discipline has taken place only within the last hundred years or so. However, the remote beginnings go back to ancient times as part and parcel of the growth and development of the all-embracing discipline of soil science. Approaches to the scientific study of soil, especially the aspects dealing with soil classification, have changed over the years as man's concept of what the soil is has changed; as the accumulation of soil knowledge has increased; and as this knowledge has been applied more and more to the solution of practical problems (for example, in crop production and land-use planning). As with other sciences there are two forces which have led to the rise and development of the discipline: the importance of *use* or relevance; and the quest after *abstract knowledge* to discover 'the laws of nature'.

In its earliest stages people became interested in studying the soil because of its importance to plant and crop growth. They became conscious of differences between soils and could differentiate between 'good' and 'bad' soils depending on the performance of their crops. Even today in many parts of the world, much soil knowledge is still of this simple but practical, utilitarian nature. The peasant's grouping of soils into 'good' and 'bad' is based on a fair knowledge of local conditions and experience gained in the use of the resource. He is able to assess soil capability through its colour, its texture, its topographic position, the species of plants growing in the area, and so on. However, for a long time farmers and scientists alike were unable to understand the reasons for the differences in the ability of soils to support crop production. They marvelled at the mystery of soil–plant relations.

The chemists were among the first scientists to study soils, following the development of the theory of mineral cycling, first in the seventeenth century but particularly in the nineteenth century. This theory, applied to soils, was at one time thought to be the answer to the age-long problem of soil fertility. Accordingly, soil science became a part of chemistry and soils began to be presented in terms of chemical elements and processes. Soil was considered either as a chemical laboratory where various chemical processes and reactions take place or as a test tube into which plant nutrients can be introduced for the benefit of plants. Soils were also classified on the basis of their chemical character or specially noticeable constituents, e.g. lime, marl, or sulphate soils. The chemical-laboratory concept is still conspicuous in the approach of some soil scientists today.

Other physical scientists, for instance physicists and geologists, also made their own contributions. Among other things, they presented the soil as a collection of matter—solid, liquid, and gas—having recognizable mechanical properties in relation to the rocks from which it was formed. To them, therefore, physical properties distinguished soils from the underlying material, as well as determining the capability of soils for crop production. For a long time many scholars regarded the soil as a geological material, the product of weathering. Thus soils were differentiated as *granite soils, diorite soils, basalt soils*, and so on, according to the types of rocks from which they were formed. That the influence of this school of soil science was strong and long-lasting is attested to by the number of present-day scientists who define soil mainly as a collection of loose debris at the earth's surface.

Thus, up till the late nineteenth century, scientists were not agreed on the definition and scope of the subject. Textbooks written on soils, even those having the same title, differed widely in the subject matter treated, so that the views held then of soils had some fundamental shortcomings. The study and classification of soils were based on parameters (geological, chemical) external to the soil and not on soil characteristics themselves. If one were to regard soil merely as an ordinary geological formation, as a crop producer, or as a nourishing layer for plants, there would be no uniform basis on which to establish a systematic scheme of classification. There was clearly a need to change these narrow ideas about the soil, to regard it as an entity on its own whose development and distribution are governed by factors which obey general rather than local laws. The formulation of these new ideas was due largely to the Russian school of pedology.

The Russian period of soil science is that of V. V. Dokuchaev (1846–1903), Nikolai Sibirtzev (1860–99), and their contemporaries and pupils. This period covers almost a hundred years, from 1840 to 1930. Dokuchaev was a geologist by training, and was for many years the head of the Faculty of Mineralogy at St Petersburg University. He devoted the greater part of his active life to the study of Russian soils and the development of an organized science called pedology. The immediate cause of Dokuchaev's interest in soils seems to be what was then known in Russia as the 'chernozem problem'—the difficulties which around 1870 were facing Russian farmers in the chernozem region.

Accordingly, the first important study made by Dokuchaev was undertaken between 1871 and 1881, under the auspices of the Russian 'Chernozem Commission of the Free Economic Society'. The result was published in a monograph with the title 'Russian Chernozem', in 1883, as a doctoral dissertation. Among other things, Dokuchaev showed for the first time that soils form distinct natural kingdoms which, like other similar kingdoms such as minerals, vegetables, and animal organisms, can be distinguished on the basis of their intrinsic properties, processes of formation (origin), and patterns of distribution. Like any other natural body, the soil has 'its past, its life, and its genesis'. Dokuchaev discovered that soils have formed as a result of the interaction of a number of factors and forces (processes), most of which can be identified and studied either separately or together in their interaction with one another. He particularly singled out certain environmental factors, including rock type, relief or topography, climate, and plant and animal life, which, acting together over a tolerably long period of time, will produce soils according to the nature and influence of the various factors. Dokuchaev had shown for the first time, in effect, that soil is both a part and a product of the environment in which it occurs and so cannot be properly understood outside it.

Having established all these concepts and principles about soils, Dokuchaev then went on to apply them to the study of Russian soils. He described standard procedures and methods for soil study, soil classification, and soil mapping. He also proposed a number of natural laws in respect of soils, the most important and subsequently the most commonly applied of which is that of zonality of soils, basic to our present-day approach to geographical soil study. Finally, Dokuchaev in his many published monographs organized the study of soils as a scientific discipline. He therefore ranks equal with Darwin in biology, Clement in vegetation studies, Davis in geomorphology, and Lyell in geology, while the Russian period roughly coincides in time with periods of similar marked shifts in the content and methods of study of a number of other natural bodies; for example, biology and Darwin's evolutionary concept; geomorphology and the Davisian-cycle concept; Clement and the climax concept in vegetation study; and climatology and Koppen's worldwide systematic classification of climates. Dokuchaev's work clearly marked the turn of a new era in soil science from that aptly described by Joffe (1939, p. 19) as the period of groping in the dark over the position of the soil in the scheme of natural phenomena.

The immediate effect of Dokuchaev's works and ideas was initially limited to Soviet Russia, where his contemporaries and pupils undertook many soil survey and research projects. For example, Nikolai Sibirtzev was responsible for the first comprehensive genetic classification of soils based on the zonal concept. These men also started organized courses in pedology, leading to a marked increase in the number of both academic and practical (field) soil scientists. The activities of some of these Russian scientists were described among others, by Vilenskii (1957, pp. 18–33), who also listed the characteristic features of the Russian (Dokuchaev) period as follows (p. 27):

1 recognition of the chief properties of soils both as bodies in nature and as the means of creating fertility;
2 abandonment of concepts of soil as simply a product of the geological processes of weathering and the recognition of the important role played by complex biological and biochemical processes in forming soil and in determining soil fertility;

3 the introduction of the evolutionary approach and the genetic method into soil study;
4 detailed study of soil-forming processes as well as of soil material using various petrographical techniques;
5 creation of soil research institutions, including the Soil Institute im Dokuchaev of the USSR Academy of Sciences and the Agricultural Academy im Timiryazev;
6 close linking of pedology with productivity and the intensification of a number of specialist studies of the soil material.

The year 1927 marks another important signpost in the development of soil science. In that year C. F. Marbut (1863–1935), a geographer, translated some of Dokuchaev's publications into English, and the effect on the Western world in particular was tremendous. Among other things the translation led to the adoption of the Russian ideas and methods in place of the prevailing agrogeologic viewpoint. This ushered in a period of worldwide compatibility of concepts, methods, and results of soils study leading to the standardization of terminology, nomenclatures, mapping procedure, and so on.

The American school of pedology led by such men as M. Whitney and C. F. Marbut must therefore be given credit for introducing and communicating Russian terms and concepts to the English-speaking world at least. Russian terms and concepts were disseminated through, among other media, the *Journal of Soil Sciences* which was first issued in 1916, and more so through the *Proceedings* of the *Soil Science Society of America*, from 1936 onwards. The Americans also made their own independent contributions to the development of pedology especially in the fields of soil survey and soil classification. They introduced and defined the *soil type* and the *soil series* as units of soil classification. The United States *Soil Survey Manual*, which is now widely in use throughout the world, was published in 1951. It describes methods of soil survey and gives guidelines on how soil profile characteristics are to be measured and recorded in the field.

The years since the Second World War have seen further improvement of the Russian legacy. International co-operation in soil study programmes and the development of experimental pedology are only two of the latest developments or improvements. In the former case, we may mention the activities of international organizations such as the Food and Agricultural Organization (FAO) and the many international congresses which meet regularly at different places the world over. The latter development seeks to correct the idea that soil study is mainly a natural field observation science not susceptible to experimental investigation. Now the various processes and factors of soil formation are being simulated, controlled, and subjected to experiment within standard laboratories.

Finally, soil is now recognized as a part of man's total physical environment. The pedologist is now becoming aware of the fact that although soils are autonomous natural bodies, they are also closely linked with the other natural phenomena. The latest phase therefore is that of comprehensive and interdisciplinary investigation of man's total environment, which means that the contribution of pedology to the solution of man's problems will be assessed, more than ever before, in terms of its use in teamwork and multidisciplinary studies than as pedology *per se*. In short, pedology, like a number of other natural sciences, is now emphasizing the relevance, as well as the nature, of the relationship between soils and the other elements and systems of the earth's surface as the home of man.

Definition and Study of Soil

The question 'what is soil?' has still to be answered. The discussion so far has been intended to show the extent of the problem of precise definition. As long as people looked at the soil from a biased viewpoint, an acceptable definition was difficult. Thus definitions such as 'soil is what plants grow on', 'the loose materials which support vegetation', 'a mixture of solid, liquid, and gaseous materials which may support plant growth', or 'the more or less loose and friable material in which plants find foothold, nourishment, and other conditions of growth', all illustrative of the agronomic concept, are incomplete and are not likely to be acceptable to the other equally important viewpoints. Other biased definitions such as 'soil is the uppermost layer of the solid crust of the earth consisting of decomposed and chemically altered rock and biotic materials' will be similarly incomplete. This last definition does not present the soil as different from saprolite, nor does it show the soil as having some individuality, resulting from the operation of specific processes and factors.

By contrast, Dokuchaev defined the soil as 'the surface and adjoining horizons of parent material which have undergone, more or less, a natural change under the influence of water, air, and various species of organisms living or dead; this change is reflected, to a certain degree, in the composition, structure, and colour of the products of weathering.' This is obviously a considerable improvement on previous definitions, although it fails to make mention of the morphologic characteristics, or the functions of soils. Marbut (1927) tried to improve on Dokuchaev's definition by mentioning the factors of reaction processes and morphology, while Joffe (1949) defined soil as a natural body of mineral and organic constituents, differentiated into horizons of variable depth, which differs from the material below in morphology, physical makeup, chemical properties and composition, and biological characteristics.

These definitions are all deficient in one way or another. Marbut fails to mention the fact that soil is a natural body, an important point especially as far as pedology is concerned. The phrase 'outer layer of the earth's crust' also precludes buried soils from his definition. Joffe's definition is apparently the most comprehensive ever written, but leaves out a number of important aspects such as soil nutrient and plant growth. Soil may therefore be defined as *an autonomous (not independent) natural body of mineral, organic matter, and nutrient constituents, resulting from the interaction of the country rock with environmental factors of climate, topography, plant and animal life, and differing from the underlying material in morphology (appearance); physical, chemical and mineralogical characteristics; and the way and manner in which it supports and reacts to plant growth.* A course in pedology should therefore include aspects of soil morphology, genesis, constituents, properties, distribution, utilization, conservation, and management.

A brief review of the courses and textbooks currently being used shows that many of these are left out. The general practice seems to be to describe: (a) the soil profile morphologically; (b) the various factors (usually rock type, climate, topography, biota, and time), and processes (podzolization, ferruginization, calcification, etc.) of soil formation; and (c) a number of major (global) soil types in relation to their characteristic bioclimatic environments (cf. Joffe, 1949; Vilenskii, 1950; Gerasimov and Glazovskaya, 1965; Bunting, 1965). In addition to these aspects, it is desirable to describe: (d) the properties and constituents of soil so as to facilitate their recognition; (e) the methods and techniques of soil survey and soil study, especially for purposes of standardization and the imparting of certain skills; (f) soil erosion and soil conservation problems, especially the role of man in the formation, destruction, and use of soils; and (g) some interdisciplinary (multiple-purpose) projects and methods. This is definitely an extensive programme, but it is also one which must be accomplished if we are to understand soils, realize the maximum benefits from their study, and appreciate intelligently the place of soil in the total picture of ecosystem communities comprising biotic and abiotic systems in mutual interaction. This book is aimed at achieving these goals.

Chapter 2

SOIL CONSTITUENTS AND PROPERTIES

The soil is a natural body with characteristic horizontal pedo-genetic layers found on the surface of the earth, which has been physically, chemically, and biochemically weathered and altered from the original bedrock or waste mantle from which it was formed. The physical and chemical processes involved in its formation are induced by climatic and hydrologic factors and the biochemical processes by micro-organisms and organic matter in the soil. Thus soil material or soil constituents derive from four main sources: the lithosphere (rocks); the atmosphere (air); the hydrosphere (water); and the biosphere (living organisms). The complex nature of the soil, its spatial distribution, and behaviour under different uses are due to the interplay of these various factors.

Soil Constituents

The soil body may be regarded in the first place as being made up of two main components: solid materials and pore spaces. The solid component is made up of (1) mineral matter and (2) organic matter. Pore spaces are of two types: (1) macro-pores, which are spaces between soil aggregates and which are usually filled by air, and (2) micro-pores, smaller

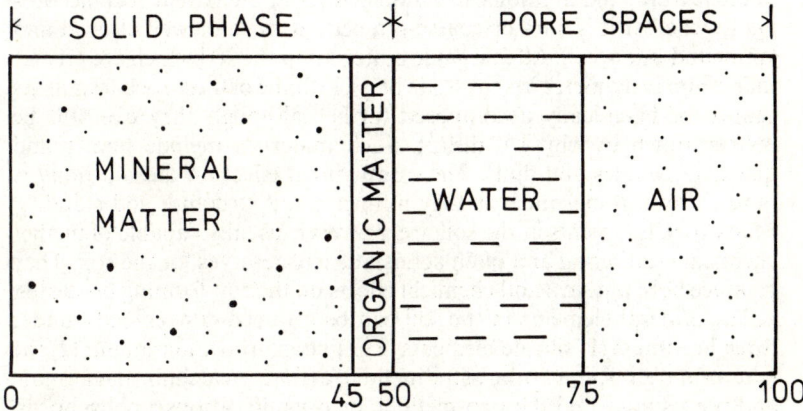

Fig. 2.1   Volume composition of a silty-loam surface soil when in ideal condition for plant growth

spaces between soil particles within soil aggregates, which are usually occupied by water. Mineral matter, organic matter, air, and water are thus the four major components of the soil. The volume composition, that is, the proportion by volume of each of the soil constituents, varies between soil types. Figure 2.1 shows the volume composition of a silty-loam topsoil which is well drained, well aerated, and ideal for plant growth. The solid state and the pore spaces each make up 50 per cent of the soil volume distributed as follows: mineral matter 45 per cent, organic matter 5 per cent, macro-pores 25 per cent and micro-pores 25 per cent. Soil aeration and drainage depend on the relative proportions of the solid materials and the pore spaces and, more especially, on the ratio of micro- to macro-pores.

For practical purposes and in order to aid the understanding of their character and significance in the soil, the components are treated separately. But one should bear in mind the fact that they are normally very well intermixed in the soil.

Mineral Matter
The mineral materials include all inorganic substances in the soil. They are the rock fragments, and the primary and secondary minerals consisting, on the one hand, of the soil particles sand, silt, and clay, and, on the other, of specific mineral elements in the soil which result from the chemical decomposition of the original rock material.

The rock fragments are undecomposed remnants of solid rock found within the soil body. The stoniness of the soil is determined by the frequency or abundance of these rock fragments. The abundance of rock fragments in the soil, the sizes and shapes of the rocks, are all related to the textural and lithological characteristics of the parent rock according to whether it is fine- or coarse-grained, massive or well cleaved and laminated, and so on. All inorganic materials in the soil which are greater than 2 mm in diameter pass as stone or rock. Some of these rock fragments cannot be chemically decomposed further although they can still be broken down by physical means. Such materials include quartz and quartzite pebbles and flints. Their only direct effect on the soil body is purely physical inasmuch as they influence soil structure and stability. Many rock fragments in the soil are, however, usually capable of further chemical weathering and often act as nutrient reserves for the soil. They produce both physical and chemical effects on the soil-forming processes.

The mineral elements in the soil can be grouped conveniently under three headings: (1) silicate elements, e.g. silicon, iron, aluminium; (2) the macro-nutrients, consisting of the metallic cations—calcium, magnesium, and potassium—and the non-metallic or organic cations—phosphorus, nitrogen, and sulphur; (3) the micro-nutrients or trace elements like manganese, molybdenum, titanium, copper, lead, zinc, and vanadium.

The non-metallic cations are discussed more fully below in the section dealing with soil organic matter.

The silicate minerals are the most abundant in rocks; therefore they form the largest proportion of mineral elements found in the soil. Their relative abundance in the soil varies with the mineralogical composition of the rock formations. The silicate minerals are derived from such rock-forming minerals as micas and feldspars and they are usually found in the more acidic rocks such as granite, shale, sandstones, and gritstones. Silicon, iron, and aluminium combine with oxygen to form oxides. Quartz (SiO_2), the oxide of silicon, is the commonest and most resistant element in soils. It is not soluble in most natural acids. The quartz grains vary widely in size from very fine particles to large pebbly fragments. Iron and aluminium can exist in both hydrated and dehydrated forms. They are found as coatings on soil particles, and within humus particles, as cement between soil particles, or as discrete particles in the soil. In conditions of low acidity iron and aluminium are stable but they become mobile in conditions of extreme acidity when they may poison plants and interfere with their uptake of nutrients.

The metallic cations are derived from rocks. Calcium is the chief exchangeable base in the soil. It is derived mainly from calcium carbonate but also from calcium sulphate. Its abundance in the soil is the main determinant of soil reaction and nutrient status. Unfortunately, it is easily leached and washed away by drainage water. Magnesium is about the second most abundant exchangeable base in the soil and it is an essential ingredient for plant growth. It is derived chiefly from ferromagnesian rock-forming minerals such as olivine, hornblende, and pyroxene. It is also associated with calcium carbonate in magnesian limestones. Potassium is found in silicate minerals such as orthoclase and potash micas. It exists in two forms in the soil: the readily available and the relatively unavailable forms. The readily available potassium is found in the soil solution and as an exchangeable base on the negative exchange sites on clays. The greatest proportion of soil potassium is found in mineral compounds such as feldspars and micas in which state it is unavailable to plants. Another metallic cation which is found in the soil but is not a major nutrient is sodium. It is the least common of the metallic cations in soils but is an important element in alkali soils of arid lands.

Sand is made up mostly of the mineral quartz (silica, SiO_2). However, the soil sand fraction, in its widest definition, includes all inorganic soil particles, whether quartz or not, whose diameters range between 0·02 and 2·0 millimetres. Sand particles are chemically inactive and therefore cannot hold nutrient cations nor enter into atomic bondage with other mineral elements in the soil. This fact, coupled with their comparatively large sizes and irregular shapes, accounts for the rather loose, granular structure of sandy soils and their high porosity and permeability. Soils with more than 70 per cent sand in their composition are usually referred to as sandy soils. Because water moves through so freely, sandy soils are subject to intense leaching and are more susceptible to drought than any other type of soil.

The free movement of water through sand is mainly downwards. Upward movement of water by capillarity above the water table is very limited. This is due to the great quantity of large pore spaces which reduce tensional forces acting on the water relative to the force of gravity tending to promote downward movement. Sandy soils also have very little ability to retain fertilizers in the soil. However, their loose structure and permeability to air and water make sandy soils particularly suitable for the cultivation of certain types of crops, notably legumes, especially in frosty and wintry environments.

Silt particles are chemically and mineralogically similar to sand, but they are smaller in size (0·02–0·002 mm in diameter), and contain a greater proportion of secondary minerals in their composition. The importance of the silt particles lies in their influence on soil structure and pore space distribution in the soil. Fine silt particles are apt to block the pore spaces between soil aggregates (macro-pores), thereby causing poor drainage and anaerobic conditions. Highly silty soils are liable to clod when wet and become compact when dry. Silt has limited permeability but the rate of capillary rise of groundwater through it is high.

Clay is the most fundamentally important of the soil mineral constituents, composed entirely of chemically active secondary minerals formed by the chemical decomposition of primary rock minerals. A knowledge of the clay minerals is fundamental to an understanding of soil chemistry and soil fertility. Clay particles are largely microscopic, being less than 0·002 mm in diameter. The study of the structure of the clay minerals had to await the discovery of X-ray photography. The formation, structure, and chemistry of the clay minerals are discussed later in this chapter. Clay soils are heavy. They are plastic and sticky when wet but they harden and crack on drying. Some clays swell when wetted but shrink on drying, thus becoming a hard and compact mass. Many of the properties of clay arise from the microscopic size of the particles which increases their volume and total surface area and affects the ability of air, water, and soil colloids to move freely within the soil.

Organic Matter

The term 'soil organic matter' refers to all dead, freshly fallen, decomposed, or only partly decomposed plant and animal materials in and on the soil. In a much wider sense the term also embraces the community of living things—micro- and macro-organisms—in the soil. The primary source of soil organic matter is plant tissue. Animals are usually considered as secondary sources.

Plant materials are composed of a number of substances of which the most important are lignin, cellulose, proteins, fats, and waxes. Lignin is the woody part of plants and the most resistant to decomposition. It is broken down, albeit slowly, by micro-organisms and by such chemical processes as oxidation and hydrolysis. Protein is rapidly decomposed releasing plant nutrients, carbon dioxide, water, and ammonia which can be oxidized to form nitrates.

When plants and animals die their remains or litter are decomposed by the living organisms to form humus and distributed by various means in the soil profile. Humus is a dark, structureless, and gelatinous substance. Very little is known about its chemical composition. It constitutes the true soil organic matter and it plays many important roles in the soil. These are discussed later in this chapter.

The rate at which decomposition of organic litter takes place depends on such factors as the balance between rainfall and evapotranspiration, temperature, ground drainage conditions, and the nutrient status of the plant material. In warm and humid tropical environments the rate of decomposition is very high but this is not so in cool or cold temperate latitudes. In temperate latitudes, low temperatures coupled with high precipitation account for the accumulation and persistence of plant materials in forest floors. Low rates of evapotranspiration and poor drainage conditions reduce the rate of decomposition. Under such conditions bacteria and other micro-organisms cannot thrive and so plant residues are not readily broken down, nor are they well decomposed. Depending on the stage of decomposition and the physical relationship with soil mineral particles, it is possible to recognize different types of soil organic matter. Raw or partly decomposed organic matter that is not intimately mixed with the mineral soil is known as mor humus. It usually consists of three layers: L (litter), F (fermentation), and H (humic, amorphous). The well-decomposed, finely divided organic matter which is intimately mixed with soil inorganic material is known as mull humus. 'Peat' refers to a soil or soil layer composed almost entirely of decayed vegetation. Peat is found most commonly in the cool and cold temperate regions, especially in bogs and poorly drained alluvial or fluvio-glacial terraces and broad plateau surfaces. But it is also found in tropical regions, especially in poorly drained, low-lying coastal river valleys and marshlands where water stagnates on the surface. For example, thick beds of peat can be found in the broad river valleys, creeks, and swampy depressions along the coast of Nigeria.

The living organisms in the soil include: micro-flora such as bacteria, actinomycetes, fungi, algae, and lichens; micro-fauna such as protozoa (e.g. amoeba and other ciliates); and meso- and macro-fauna including nematodes, ants, termites, eelworms, earthworms, molluscs, and arthropods. Tree roots also form part of the community of living things in the soil.

The importance of these organisms in the soil derives from the way in which they obtain energy for their metabolism. Algae and some bacteria obtain energy from the sun and the atmosphere. In the process they help in fixing atmospheric nitrogen and adding it to the soil. In the same manner they obtain oxygen and carbon from atmospheric sources. By contrast, lichens and fungi are capable of obtaining nutrients from rock minerals by ion exchange, while the vast majority, especially the soil fauna, obtain their energy by digesting and oxidizing, through enzymes, organic matter in the soil. Plant debris is broken down and converted into carbon dioxide and water and at the same time nitrogen and other organic nutrients are released.

Generally, soil organisms help in building soil structure while the larger animals improve soil aeration by burrowing.

Soil Air

As has been pointed out, there are pore spaces between soil particles within soil aggregates, as between the soil aggregates themselves. The spaces are generally either triangular or rhomboidal in shape. The pore spaces between soil aggregates, the macro-pores, are normally filled by air. In practice it is the amount of water in the soil which determines the volume of air. Soil air constitutes the soil atmosphere from which plants and soil organisms obtain oxygen for their metabolism, and dispose of carbon dioxide and other obnoxious gases. Its composition therefore varies according to the rate at which organisms take up oxygen and deposit carbon dioxide and the rate at which it is replenished from the atmosphere.

The carbon dioxide content is not only highly variable compared to that of the atmosphere, it is also considerably higher. The carbon dioxide in the soil atmosphere comes from plant roots, micro-organisms, decaying organic matter, and carbonic acid in rain and drainage water. Unlike the atmosphere, however, the soil air is always saturated with water vapour (Table 2.1). The oxygen content of the soil air is somewhat lower and more variable than that of the atmosphere. The oxygen content of the atmosphere is about 21 per cent by volume while that of the soil air varies between 15 and 20 per cent. Argon is another gas which is found both in the atmosphere and in soil air in very small amounts, which do not differ much between the two media. Only a very small proportion of soil nitrogen is contained in the soil air although it makes up about 79 per cent by weight of the atmosphere.

The diffusion of air in the soil is through the macro-pores and is greatly facilitated by cracks, fissures, and animal burrows. The rate of diffusion is thought to be proportional to the square of the air space (macro-pores) in the soil. In well aerated soils, micro-organisms can thrive, thus enhancing a vigorous and rapid rate of biological activity. Root hairs, through

SOIL CONSTITUENTS AND PROPERTIES

TABLE 2.1
Composition of Soil Air Compared to the Atmosphere

	Relative humidity	CO_2	O_2	N_2
Soil air	100%	Variable 2–20%	Variable 15–20%	Variable 0·02–0·4%
Atmosphere	Variable	0·03%	21%	79%

which water and nutrients pass from the soil into plants, can only function efficiently in the presence of oxygen in the soil. In the absence of oxygen, for example in a waterlogged soil, substances toxic to plants may accumulate. In this anaerobic condition soil acidity increases and this may cause the dissolution of certain mineral elements, such as iron and aluminium, which then interfere with plant uptake of essential nutrients. Oxidation and reduction processes are also related to soil aeration. These points are discussed in greater detail in Chapter 9 which deals with soil–plant relations.

Soil Water
Soil water, together with its dissolved salts, constitutes the soil solution, the medium through which many chemical processes take place in the soil and plants are supplied with nutrients. The soil receives most of its water supply from the local rainfall. When rain falls part of the water sinks into the soil. The rate at which it sinks through the soil, that is, the soil infiltration capacity and permeability, depends on the texture, structure, and initial moisture content of the soil.

Water is held in the soil in the micro-pores between soil particles and on the surfaces of the clay and organic matter particles. They are called capillary and adsorbed, or, hygroscopic water respectively. A third type of soil water is the gravitational or free-draining water. This is the water which occupies the macro-pores, commonly after each rainfall, moving through the soil to the water table under the force of gravity. When all the micro- and macro-pores are filled with water the soil is said to be waterlogged. The soil is at field capacity when all the macro-pores are empty but the micro-pores are completely filled with water.

Gravitational water is not of any importance to plant growth and in fact may constitute a hazard in that when the macro-pores are filled with water, air is expelled and plant roots and micro-organisms suffer from lack of oxygen. Its importance is in leaching and eluviation processes in the soil.

Adsorbed and capillary water are sometimes called 'immovable' water as against the 'movable' free-draining or gravitational water. In adsorbed water, the molecules are held by electric charges on the clay and organic-matter particles. Capillary water is held in the micro-pores by a force greater than the force of gravity and this force is inversely proportional to the size and shape of the micro-pores. The critical pore size is thought to be an equivalent diameter of about 30 μ. Capillary water was once thought to represent water moving up towards the soil surface from the water table. But it is now known that capillary rise of water from the water table is limited to a short distance above it, to the area known as the capillary fringe.

There will be a constant supply of water to plants as long as the amount in the soil is adequate. As the soil dries out it becomes increasingly difficult for plants to obtain water until a point is reached where the plants begin to wilt. This point at which the water supply to plants is so small that they wilt is known as the wilting point and the soil moisture content at that stage is referred to as the wilting coefficient. The percentage of soil water remaining after the plants have become permanently wilted is referred to as the permanent wilting percentage of the soil. This water is usually held in thin films in the smaller micro-pores and on the clay and humus particles. Even at its driest state, when all the micro-pores are devoid of water, some water is still left in the soil and this is held tightly and immovably on the clay surfaces. The amount of water in the soil at this stage constitutes the hygroscopic coefficient of the soil. Capillary water is often defined as that held in the soil between field capacity and the hygroscopic coefficient.

Soil water is lost to the atmosphere by direct evaporation from bare surfaces and by transpiration in vegetated areas. The rate of evapotranspiration depends on atmospheric temperature and relative humidity and also on soil temperature and moisture content. The rate is higher in a warm or moist than in a cool or dry soil. As the soil dries out the tenacity with which soil water is held increases, thus slowing down the rate of loss to the atmosphere. Plants transpire most efficiently when the soil is near or at field capacity, when their metabolic rate is at its best. If the soil is waterlogged the lack of oxygen supply to the roots greatly reduces the rate at which they can take up water and transpire.

Water is also lost to the soil by percolation down and out of the soil profile. Percolating water is important in the eluviation and leaching processes in the soil. Fine mineral, salt, and organic matter particles are physically washed away. The acidic nature of percolating water makes it a potent weathering agent dissolving soluble salts and mobilizing nutrient cations and soil colloids. The intensity of eluviation and leaching depends, of course, on the amount of water passing through the soil which is in turn related to the amount of rainfall and the soil texture and porosity.

Soil Properties

Thus far an attempt has been made to describe the major constituents of the soil body. Each of these constituents performs important functions in the soil and they interact with one another to give the soil certain distinctive properties or characteristics. These properties fall into two major groups: the physical and the chemical (including biochemical) properties. The physical properties include colour, texture, structure, porosity, temperature, etc., while the chemical or biochemical properties include cation exchange capacity, exchangeable bases, soil reaction, organic matter content, and so on. These will now be discussed.

Soil Colour

Colour is the most conspicuous property of soil and has been widely used in its description and classification. Soils are commonly described as (1) dark, in which case they may be black, dark grey, dark brown, or cinnamon; (2) bright, e.g. yellow, orange, red, reddish-brown, yellow-brown, etc.; or (3) light, e.g. white, whitish-grey, etc. These colours may predominate at certain portions of the soil profile, for instance, the generally dark humus (Ao) layer, the light (ash-grey) eluvial layer in a typical podzol, and the bright illuvial or B layer also in a podzol profile. In some cases, however, these colours are mixed, as in mottled soils, or as a result of a mixture of the various colour-producing elements which may then produce a diversity of colours and tinges.

Two main substances produce colours in soils: organic matter and mineral elements. In general, organic matter, particularly decayed plants or humus, produces the dark colours, while various inorganic (mineral) substances produce different hues and shades of colour. The oxides of iron (ferric iron) and manganese give red, brown, and yellow colours, while ferrous iron gives a bluish tinge. Similarly, lime tends to intensify the dark colours, e.g. of humus, mica gives a glittering appearance, and silica (quartz) a whitish or greyish-white colour. The colour of minerals or compounds also reflects the weathering environment in which they occur. Thus compounds in the well-drained top portions of the soil profile are generally bright brown, red, or yellow, especially if they are oxidized. By contrast, those found in saturated or reducing environments are generally dull grey. Indeed, there is a sizeable group of scientists who believe that soil elements and their proportions can be estimated accurately from their colours. This is particularly so with the oxides, especially those of iron and manganese.

Soil colour, however, may create problems. In the first place, there may be little or no direct relationship between colours and the listed elements and compounds. For instance, contrary to the general assumption, many black soils are deficient in organic matter. The colouring oxides and colloids also have different colours in different weathering stages and environments. Iron or manganese may be colourless, green, blue, yellow, brown, or red in different areas or even the same area. Finally a number of problems arise when determining soil colours. The colour of a soil when it is wet is different from its colour when it is dry. Moreover, personal idiosyncrasies may affect colour description, e.g. between male and female observers, and also more seriously, with people who are colour blind. The terminology of soil colour may also create problems if it is not standardized. Soil colour designations suffer from having no sound quantitative basis. The Munsell Color Chart is now widely used in order, at least, to infuse some objectivity into soil colour determination and nomenclature and enhance comparability and, even, statistical treatment of results.

Soil Texture
The mineral materials form the 'skeleton' or the framework of the soil. Soil texture refers to the coarseness or fineness of these materials. Soil texture is also commonly defined as the relative proportions of sand, silt, and clay in the soil. This latter definition is rather narrow and is considered by some to be unsatisfactory because it excludes particles of gravel calibre, that is, those bigger than 2 mm in diameter. As has been indicated above, particles of gravel calibre are important in soils. They are particularly prominent, for example, in soils formed from pisolitic duricrusts (laterites, ferricretes, etc.). Gravel-size particles should therefore always be considered in the analysis and classification of soils.

Soil texture is the most permanent and most fundamental soil property; it is changed little by man through soil management practices. It has considerable influence on soil structure, consistence, degree of compaction and stability, soil drainage, soil aeration, and root penetration. It determines the ability of the soil to hold and exchange nutrients and it is a crucial factor in determining soil response to liming and fertilizer applications on agricultural lands.

Soil textural grades are defined on the basis of the size of particles. The size of a particle is usually the diameter of a circle described by its longest axis. Unfortunately different limits are set by different countries or organizations to define the size grades as shown in Fig. 2.2. The various systems—European, American, International, etc.—have their advantages and disadvantages but the International system is preferred mainly on account of its simplicity and ease of handling. Not only does it have fewer categories or grades than the other systems, but it is metric and the values are readily convertible into other scales, e.g. the phi-scale.

Soil texture may be studied directly in the field or in the laboratory (see Chapter 6). In many cases, the field assessment is sufficient to provide the necessary information. However, when accurate and precise information is required, especially for statistical analysis, laboratory analysis is

SOIL CONSTITUENTS AND PROPERTIES

1. THE INTERNATIONAL SCALE

Clay	Silt	Sand		Gravel
		Fine	Coarse	
0.002	0.02	0.2		2

2. U S DEPARTMENT OF AGRICULTURE

Clay	Silt	Sand	Gravel
0.002	0.05		2

3. U S BUREAU OF SOILS AND PUBLIC ROADS ADMINISTRATION

Clay	Silt	Sand		Gravel
		Fine	Coarse	
0.002	0.05	0.25		2

4. MASSACHUSETTS INSTITUTE OF TECHNOLOGY AND BRITISH STANDARDS INSTITUTE

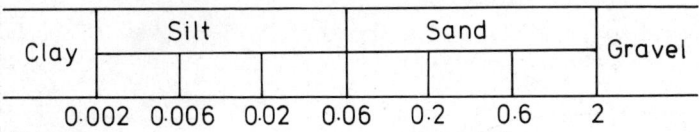

All scales are in mm.

Fig. 2.2 Systems of texture grades

essential. The determination of soil texture, as is done in the field (see Chapter 6) by rubbing a sample between the fingers, can only provide enough information to allow one to decide whether the soil is coarse, medium, or fine textured; or whether it is light or heavy. It cannot give accurate measures of the proportions accounted for by each particle-size grade. When rubbed between the fingers sand has a gritty feel, clay is usually sticky, while silt has a smooth, silky or soapy feel.

There are twelve soil textural classes on the whole as shown in Fig. 6.2. They are arranged in sequence from soils that are coarse in texture and easy to handle (sandy soils) to soils that are very fine textured and

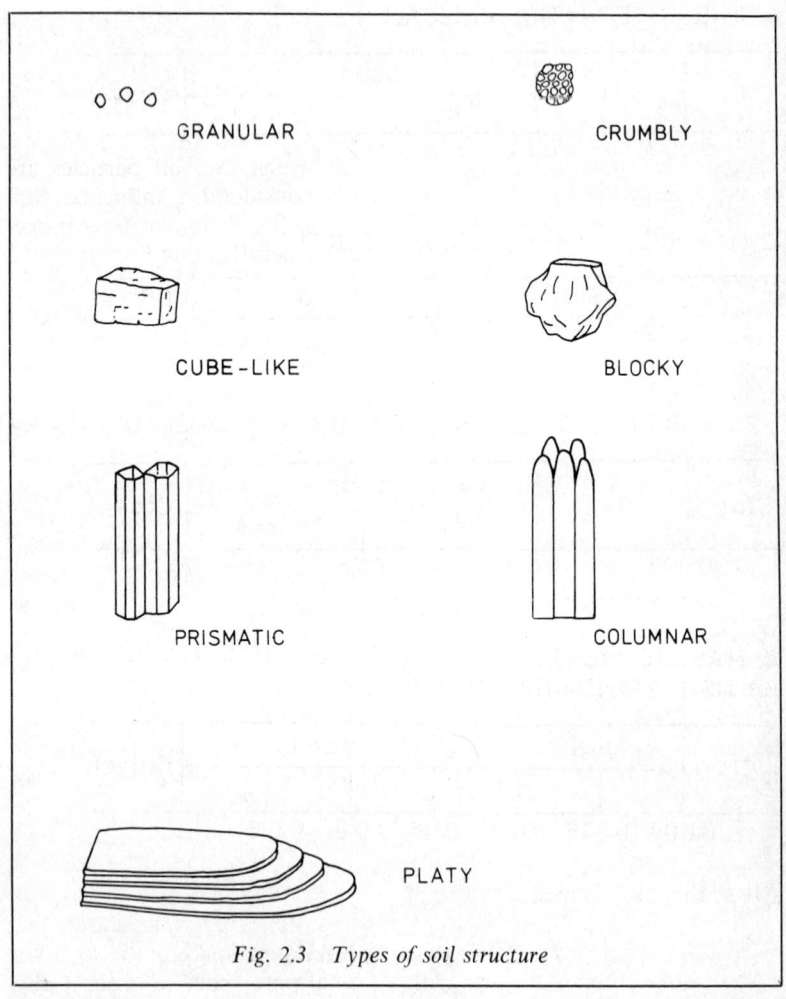

Fig. 2.3 Types of soil structure

difficult to manage (the heavy clays). Most soils can be described as either sandy, clayey, or loamy. Sandy soils are relatively easy to determine, as the sand particles can be seen and felt. The same is true of clayey soils. A soil is described as clay when the proportion of clay-size particles is at least 40 per cent. However, few soils are predominantly sand or clay; most soils come under the category of loam, comprising all the possible size grades. Loamy soils are classified according to the major component particle-size grade or grades, e.g. sandy loam, sandy clay-loam, silt-loam, silty clay-loam, etc. The definition and identification of the various soil textural classes in the field are described in detail in Chapter 6 (see in particular Table 6.3).

Soil Structure
Apart from the relative proportions of sand, silt, and clay which determine many soil characteristics, the manner in which the soil particles are grouped together in the soil also exerts considerable influence. Soil structure is a descriptive term which refers to the characteristic arrangement of the soil particles. Soil particles are generally bound or cemented together by various substances including organic matter, colloidal clay, iron, aluminium and other hydrous oxides, and by the flocculating action of certain mineral salts. The resulting aggregates are known as soil *peds*. When there is little or no aggregation of particles the soil is said to be structureless, e.g. loose sand. Soil structure is usually described in three ways: (1) types of ped, (2) size of ped, and (3) strength or resistance of ped to pressure. The types of ped are defined by their shape, whether granular, platy, blocky, prismatic, or columnar, etc. (Fig. 2.3). Peds are described as fine, medium, or coarse depending on their sizes. Peds may also be graded as structureless, weak, moderate, or strong, depending on how well formed and how resistant to pressure they are (see Chapter 6).

The factors which affect or influence the development of soil structure may be grouped into those which are due to climatic influences and those which are due to the components of the soil body itself. Alternate wetting and drying of soils has been observed to be responsible for the aggregation of soil materials into clods of varying sizes; so is alternate freezing and thawing. The soil moisture condition is also important in soil aggregation. Generally speaking, soil peds are more well formed and stable in subhumid and semi-arid soils (e.g. chernozems and prairie soils) than in wet (podzols) or desert soils (sierozems). Under arid conditions, there is little clay or organic matter, while these materials are leached out of the top horizons in wet soils.

The role of organic matter in soil aggregation is very important. The addition of organic matter increases the total pore space of the soil. Not only does organic matter have a large total pore space in itself, it also increases granulation and stabilizes the aggregates. Furthermore, burrowing animals, tree roots, and tillage (cultivation) break up soil in many ways, thus encouraging the formation of peds. Calcium plays a role similar to that of organic matter. But a high sodium content as in arid or saline soils militates against ped formation.

Soil structure is difficult to measure quantitatively, but a number of methods are available for estimating it. Among these the most commonly used is that of allowing soil lumps to fall from a height and observing the shapes and sizes of the resulting fragments. Obviously, the best structure is that which increases the aeration and the water-holding capacity of the soil, and facilitates the activities of micro-organisms. Loose, friable soils are easier to cultivate than heavy, compact soils.

Soil Porosity
Porosity is commonly defined in terms of the total volume of pore or empty space between the particles of the soil material, when the soil is in its natural state. It is generally expressed as a percentage of the soil volume, given by the formula

$$P = 100\left(1 - \frac{dv}{d}\right)$$

where P = porosity, d = density or specific gravity of the soil solid material, and dv = volumetric weight of the soil, or its apparent specific gravity. The true specific gravity of most soils is 2·65 or thereabouts, and the volumetric weight of a soil can be measured easily. One other way of calculating soil porosity is to measure the volume or quantity of water or any other liquid which will completely fill up all the pore space in a given soil.

Porosity is influenced by a number of factors, including soil texture, structure, consistency, organic matter content, and biological activity. Generally speaking, the finer the soil texture the greater the total surface area and so the pore space or porosity. Consequently, clay soils have the largest percentage of pore space (50–60 per cent), and sands the lowest (20–30 per cent), while the loams are in between (30–50 per cent). By the time the mean size of the component particles of a given soil is reduced from sand (2 mm) to clay (0·002 mm), the pore space will have increased about 1,000 times. This is because sand (2 mm) particles contain on average 1,500 pores to the square centimetre as against 25,000,000 for clays.

Apart from grain size, the arrangement of the individual grains and their mode of agglomeration are also important. In general, porosity is low in well-packed (compacted) soils especially when they are made up of particles of varying sizes. Soils whose particles are cemented together, for example, by iron oxides, also have their porosity reduced. Porosity is high in granulated and in well-structured soils.

Soil Temperature
Soil temperature is a very important property of the soil as it regulates to a large extent the rate and intensity of the biochemical and chemical processes of soil formation; it is also of direct significance for plant growth. Soil temperature may be expressed in terms of its component parts, or effective factors, including specific heat, thermal conductivity, and soil reflective and radiative characteristics.

The specific heat of soils is the amount of heat required to raise the

SOIL CONSTITUENTS AND PROPERTIES

temperature of 1 g or 1 cm^3 of soil by 1°C. It depends on such factors as the specific heat of the constituent minerals of the soil and the moisture and organic matter contents. The specific heat of most soil minerals may be considered compatible in different places, particularly since the same minerals are found in most soils. However, the organic matter and moisture contents vary widely, thus making for differences in specific heat and soil temperature from place to place. Generally, the greater the organic matter content the greater the specific heat of the soil. Similarly, the more moist a soil, the greater its specific heat, because the specific heat of water is much higher than that of air. While the specific heat of water is 1, that of air is only 0·000306. Wet soils and soils rich in organic matter are therefore generally cooler than drier and less humic soils in the same climate areas, i.e. areas with similar insolation.

Soil thermal conductivity is the rate at which sensible heat is transferred from the soil surface downwards. It depends on the type of constituent minerals, the organic matter content, the moisture content, and porosity. Generally, the thermal conductivity of moist soils is higher than that of dry soils.

Finally, the nature of the soil surface considerably affects the amount of solar energy absorbed by the soil. Dark soils generally absorb more solar energy than light-coloured soils, while soils with a smooth surface reflect more radiation than those with a rough surface. Other properties of the soil which affect soil radiation characteristics are its moisture content, surface conditions, and conductivity.

Soil Drainage

Soil drainage is another important soil property and it involves a qualitative assessment of the moisture-to-air ratio of the soil, especially in relation to the needs of plants. The rate of water movement through the soil is also involved. When the soil mass is totally devoid of water, all the pore spaces are filled with air and the soil is in a fully aerobic condition. On the other hand when all the pores are filled with water the soil is waterlogged and in an anaerobic condition. Between these two extremes various moisture-to-air ratios are possible.

However, four soil drainage classes have been defined (Clarke, 1957):

(a) *Excessive drainage*, a condition in which water moves too rapidly through the soil with the result that not enough is retained for the normal growth of any plant species except those in true ecological equilibrium with it. Extremely loose sandy or stony soils are liable to be excessively drained.

(b) *Free or perfect drainage*, in which water moves easily through the soil to give ideal aeration while at the same time sufficient water is retained for normal plant or crop growth. Such soils appear moist when inspected in the field. Loamy soils are more likely to be freely drained than either the essentially sandy or clayey soils.

(c) *Imperfect or poor drainage*, which implies a fluctuation between aerobic and anaerobic conditions which may be related to seasonal fluctuations in the level of the water table or to periodic inundation of the soil by floodwater. Imperfectly drained soils may occur also at 'receiving sites' along hillslopes.

(d) *Impeded or very poor drainage*, in which there is a definite obstacle to downward percolation of water so that the soil is always saturated. Such impedance may be caused by a high water table in low-lying areas or by the presence of an impermeable layer below the surface, e.g. clay pans, laterite crust, rock pavement, etc.

Imperfect or impeded drainage usually results in 'gleying' of the soil. Gleying is the term used to describe the process which results in the 'dulling' or 'paling' of soil colours and/or mottling under an excessive moisture-to-air ratio condition. Under anaerobic conditions the action of reducing bacteria leads to the reduction of iron oxides to the ferrous state, thus changing the colour of the soil from brown, red, or yellow to grey or to the paler forms of these colours. In imperfectly drained soils some of the reduced iron will be re-oxidized to form ochres when the moisture content reduces and air fills the pores. These ochres are referred to as *mottles* whose colour may vary according to the degree of hydration. Grey mottles occur in the more imperfectly drained soils. Mottles are usually absent in soils with impeded drainage. The gley horizons of such soils tend to persist for a long time even after the land has been drained.

Clay Minerals and Soil Chemistry

Clay units, in simple language, are built of sheets of silica (SiO_2) and of hydrated alumina ($Al(OH)_3$). The silica and the alumina sheets can be combined in two ways to form the two major groups of clay minerals: the montmorillonite and the kaolin. In the montmorillonite type each alumina sheet is sandwiched between two silica sheets, the ratio of silica to aluminium being 2:1. In the kaolin type the alumina sheet lies on top of the silica sheet in the ratio 1:1 (see Fig. 2.4). In the 2:1 group when the units are joined to form the *crystal lattice* it is possible to have two silica sheets together which is not the case in the 1:1 kaolin group. Thus the bonding between the 2:1 units is very weak. By contrast, the kaolin units are held together by strong hydrogen bonds.

The clay mineral units are so thin that they can only be measured in Angstrom units (Å) or one millionth of a millimetre. The distance between corresponding sheets in adjacent units is called the base spacing. This distance is only 7 Å in the 1:1 kaolin group. In the 2:1 type the base spacing may vary between 10 A and 50 A or more. This is because

SOIL CONSTITUENTS AND PROPERTIES 25

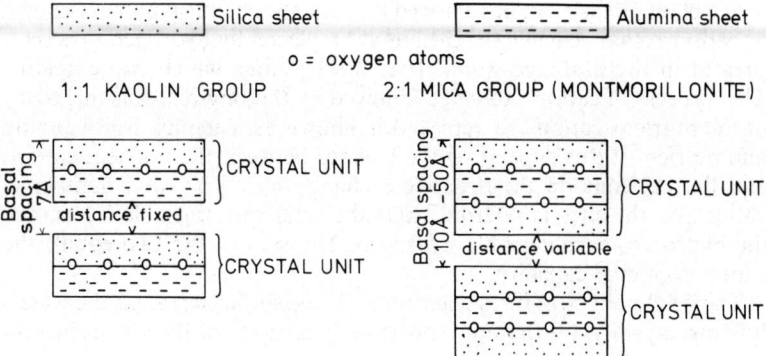

Fig. 2.4 Schematic representation of the lattice structure of the 1:1 kaolin and 2:1 montmorillonite clays

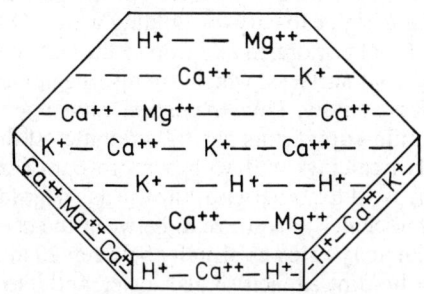

Fig. 2.5 A clay crystal with excess negative charges and adsorbed cations

water molecules can easily enter the weakly bonded spaces between the units, causing them to expand. This explains why certain types of clay swell considerably on wetting and contract again on drying.

Each clay particle is surrounded by a film of water saturated with negative electrical charges. The net negative electrical charges on the clay particles constitute part of the soil's cation exchange capacity, that is, the ability of the soil to hold nutrient cations and exchangeable hydrogen. Nutrient cations such as calcium, magnesium, and potassium are attracted to these negative exchange sites (see Fig. 2.5). Cation exchange capacity is measured in milli-equivalents per 100 grammes of dry soil.

When the cations are held on clay particles they cannot be washed away physically by drainage water. They can only be removed in either of the following ways: they can be taken up by plant roots and transmitted to the plant or they can be removed through the chemical process known as cation exchange when the cations are replaced by hydrogen ions (H^+) present in the drainage water. It is the H^+ ions which cause acidity. This process of cation exchange followed by the physical washing away of the nutrient cations so replaced is known as leaching. Soil reaction and nutrient status depend on the balance between the nutrient cations and the H^+ ions on the negative exchange sites. The total number of cations on the sites is referred to as the total exchangeable bases, and the hydrogen, exchangeable hydrogen. The sum of the two equals the cation exchange capacity.

Their different structures determine: (1) the surface area, (2) the water-holding capacity, (3) the nutrient-holding capacity of the clay minerals.

MONTMORILLONITE GROUP

Montmorillonite has a relatively simple structure and the bonding of the constituent sheets is very weak. Water molecules can readily squeeze in between the sheets causing the expansion of the particles. Montmorillonite therefore has a high capacity for holding water. On drying, it contracts and cracks. Furthermore, in addition to the outside surfaces of the particles, their internal surfaces, which are also negatively charged, are exposed to chemical activity. Thus the cation exchange capacity is very high—about 100 milli-equivalents per 100 grammes of dry soil. Another type of the 2:1 lattice clays, but with a more complex structure and stronger bonding, is illite, otherwise known as 'degraded mica'. It is harder for water molecules to squeeze in between the crystal lattice. The cation exchange capacity is lower, usually between 20 and 40 milli-equivalents. The water-holding capacity is also lower, and it expands and contracts less than montmorillonite.

KAOLIN

The crystal lattice is very compact and resistant; water molecules cannot squeeze between the sheets. Thus kaolinite has a comparatively small surface area. Its water-holding capacity is very low and kaolinite is not subject to the swelling and contracting characteristic of montmorillonite. The cation exchange capacity is of the order of 5 milli-equivalents per 100 grammes of dry soil.

A high content of kaolinite in a soil is thought to indicate an advanced stage of weathering. Montmorillonite and illite are in a secondary stage of weathering. Illite eventually weathers to kaolin by first of all being disintegrated completely and then re-forming into kaolin. Kaolin is thought to break down into silicic acid and secondary quartz as the end-products of the weathering.

SOIL CONSTITUENTS AND PROPERTIES

Role of Soil Organic Matter

As with the clay minerals, organic matter is a major soil constituent which determines many chemical and biological processes in the soil. Its role in the soil includes the following:

(1) It is a store and supplier of plant nutrients, i.e. it contains 90–95 per cent of the soil nitrogen, about 50 per cent of the phosphorus and 80 per cent of the sulphur. Nitrogen is the main source of protein in the soil and it is a constituent of chlorophyll, the green pigment of leaves by which photosynthesis takes place. Phosphorus is an important constituent of plant protoplasm. It controls the manufacture of carbohydrates during photosynthesis. But its chief role in plants is to store energy. Sulphur is also an important constituent of plants which plays a major role in their metabolism. It promotes plant growth and it is one of the elements used in manufacturing plant protein. These organic nutrients are discussed more fully in Chapter 9 on soil–plant relations.

(2) Humus shares with the clay minerals the ability to hold and exchange nutrient cations. Just as there are negative exchange sites on clay surfaces so there are on the humus. It is believed that 50 per cent of soil cation exchange capacity is made up of the negative charges on humus. Clay and humus are usually intermixed in the soil to form the so-called clay–humus complex. The clay–humus complex has 'colloidal' properties. Clay and humus in the soil are stable, that is, they retain their solid state as long as soil acidity is low. In conditions of extreme acidity they are easily dispersed in the soil solution and mobilized. Owing to their small size the particles in the colloidal solution have a very large surface area and hence high powers of adsorption.

(3) Humus particles are hygroscopic; they absorb water and are therefore important in conditioning soil moisture status.

(4) Humus contains organic polymers which help in binding together soil particles through hydrogen bonds. Humus thus helps in building soil structure.

(5) Forty to sixty per cent of humus is an acid called humic acid which takes part in weathering and leaching processes in the soil. In poorly drained conditions other types of acid solutions, such as fulvic acid, are produced which are even more powerful leaching agents. They are particularly important in the soil-forming process known as podzolization discussed later in Chapter 4.

Soil Reaction

One of the most important chemical properties of the soil is its reaction. Soil reaction refers to the acidity or alkalinity of the soil. It is so important because of its influence on many of the chemical elements and biological processes in the soil (see Chapter 9); soil reaction largely determines the chemical environment in which plants and soil micro-organisms live.

Soil acidity is the main symptom of low base status. It is caused chiefly by the concentration of H^+ ions in the soil. The H^+ ions will concentrate in the soil commonly in regions where rainfall or precipitation is high and enough water passes through the soil to leach appreciably the exchangeable bases. As has been pointed out earlier in this chapter, during cation exchange nutrient elements are removed from the negative exchange sites on the clay–humus complex, and replaced by H^+ ions. This increases their concentration in the soil solution. In general the H^+ ions come from three main sources: (1) rainwater which is a weak solution of carbonic acid, (2) organic matter, and (3) plant root exudate. Furthermore, as the H^+ ion concentration increases a stage may be reached where many aluminium compounds become soluble and are present as the trivalent Al^{3+} ions. Highly acidic soils usually have both H^+ and Al^{3+} ions. The aluminium ions are hydrolyzed to release more H^+ ions into the soil solution.

Alkalinity, which occurs in soils with high base status, results from the concentration of hydroxyl anions (OH^-) in the soil. The exchangeable bases—calcium, magnesium, and potassium—produce the hydroxyl anions upon the hydrolysis of the clay–humus complex on which they are adsorbed. The hydroxyl anions are also derived from salts such as the carbonates of calcium, magnesium, and sodium. Sodium carbonate is particularly associated with alkaline soils. The concentration of the hydroxyl anions will only take place where rainfall is not heavy enough to cause appreciable leaching of the exchangeable bases, that is, in arid, semi-arid, and subhumid regions.

Soil reaction therefore depends on the relative amounts of exchangeable hydrogen (source of H^+ ions in the soil solution) and exchangeable bases (sources of hydroxyl anions on the clay–humus complex). Where the H^+ ions predominate, acid conditions prevail, but where the OH anions predominate the soil reaction is alkaline. When the two elements are concentrated in equal amounts a condition of neutrality is established.

Soil reaction is usually determined by measuring the hydrogen ion concentration in the solution. This is done by the use of the pH meter. Soil pH is defined as the negative index of the logarithm of the hydrogen ion concentration. In other words it is the logarithm of the reciprocal of the H^+ ion concentration expressed in equation thus: $pH = \log_{10} 1/H^+$. The pH scale ranges from 1 to 14. A soil is said to be acidic if the pH is less than 7, neutral if at 7, and alkaline if it is above 7. At pH 7, as in pure distilled water, there are equal amounts (10^{-7}) of H^+ ions and OH anions. When the concentration of H^+ ions increases relative to that of OH anions soil acidity increases. For instance, at pH 6 there are ten times as many H^+ ions as there are hydroxyl anions, while at pH 5 it is a hundred and at pH 4 a thousand times. Conversely at pH 8 there are 10 times as many hydroxyl anions as there are H^+ ions, etc.

The pH range of soils is rarely beyond 3 to 10, whilst most soils have pH values between 4 and 9. The pH is, however, a very variable soil property. In the first place, generalizations cannot safely be made about the composition of a soil from pH figures alone, because pH detectors may not detect lime which occurs as nodules rather than as fine earth. Secondly, pH varies in many directions in the soil. Differences of more than 3 have been reported within a small area and in a soil classified as a single soil type. Also pH varies seasonally, in response to variations in both temperature and rainfall, as well as to differences in plant needs. The pH of stored soils also varies, depending upon the mode and the length of storage. The pH of soils stored in air-tight plastic bags is generally affected as a result of microbial activity, e.g. the occurrence of moulds. Differences of up to 2·5 have been reported in this connection. The divergence is greatest for wet soils. Therefore, pH is best measured in the field, or immediately on reaching the laboratory (see below). It is important to note that detailed, accurate laboratory determinations of pH have little or no advantage over the direct field method, e.g. the Raupach method. Finally, different methods will give different results.

The Soil Profile

So far in this chapter, an attempt has been made to describe the major constituents that make up the soil body and the physical, chemical, and biological characteristics associated with them. The embodiment of all these soil morphological characteristics is the soil profile. It is a two-dimensional cross-section of the soil. One of the most important features that distinguish the soil from ordinary weathered material is the characteristic layering of the soil profile. These layers are called horizons. Soils are often described and classified, among other things, according to the number and characteristic arrangement of horizons.

The ideal soil profile consists of three horizons—A, B, and C. The A horizon is the topmost layer and may or may not include the organic litter layer. This litter layer, where present, consists of the undecomposed, partly decomposed, and decomposed organic matter present on the soil surface. Below this layer is the eluvial (A) horizon, which is the portion of the soil profile affected by leaching or eluviation. From here have been lost most of the mobile substances supplied by the humus layer above and those released by chemical weathering in the layer itself. The materials removed from the A horizon are usually transported mechanically down the profile by percolating soil water and deposited in the B horizon, which is therefore a zone of accumulation or enrichment. It is particularly rich in the colloids of iron, aluminium, and manganese as well as in organic substances and clays, etc. Finally, below the B horizon is the zone relatively unaffected by the soil-forming (pedogenic) processes.

Fig. 2.6 The soil profile and horizon nomenclature
(a) A soil profile with organic matter accumulating on the surface
(b) A soil profile with an agricultural horizon

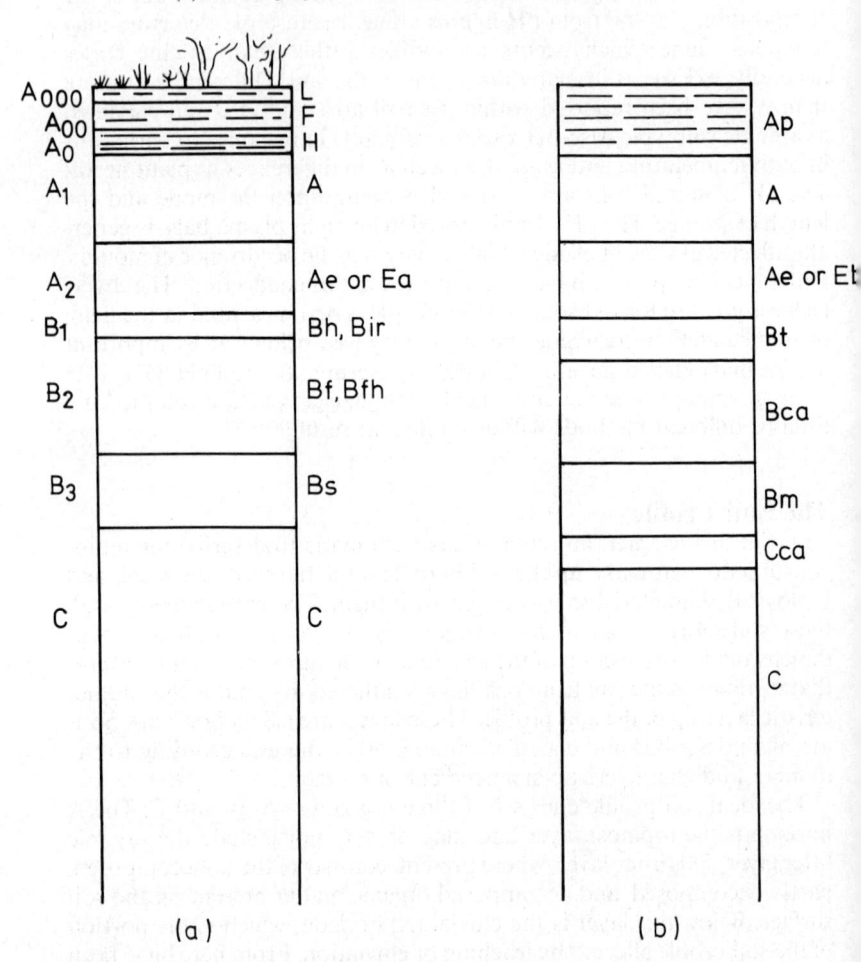

(a)		(b)
A_{000}	L	Ap
A_{00}	F	
A_0	H	
A_1	A	A
A_2	Ae or Ea	Ae or Et
B_1	Bh, Bir	Bt
B_2	Bf, Bfh	Bca
B_3	Bs	Bm
		Cca
C	C	C

This is called the C horizon or the soil parent material. Percolating water moves through the entire soil profile in many places, taking along mobile substances, clays, etc., which are then lost to underground drainage water (see Fig. 2.7).

SOIL CONSTITUENTS AND PROPERTIES

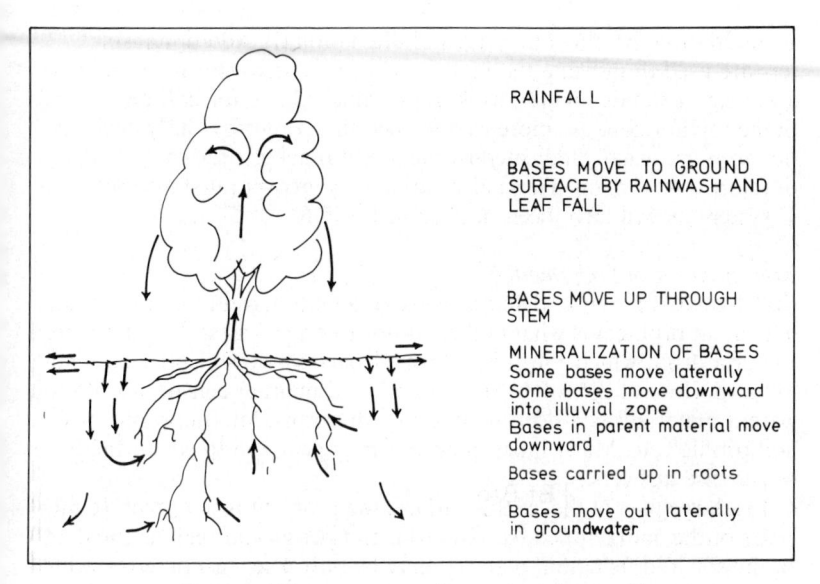

Fig. 2.7 The movement of bases in a forest soil

Originally the letters A, B, and C were introduced by Dokuchaev to label the three layers of the chernozems (Nikiforoff, 1931). A horizon was the first of the horizons from the surface; C was the parent material mainly unmodified by the soil-forming processes; while B was the transitional horizon between the two. Soon the symbols acquired a little more significance than mere labels and the A horizon, for example, became the horizon of organic matter accumulation. Later as a result of the study of podzols, chiefly by Sibirtzev and Glinka, the A horizon of podzolic soils became an 'eluvial' horizon and the B an 'illuvial' horizon. These terms are now applied to the A and B horizons of all soils regardless of their genetic type. The layer of humus accumulation, among others, has also been separated from the strictly eluvial layer; the subdivisions of horizons being described include A_1, A_2, A_3, B_1, B_2, B_3, etc. At first the subscripts indicated only the succession of the subhorizons but, currently, some of them have definite pedological significance; e.g. A_2 in all podzolic soils is the horizon of maximum leaching.

The problem with the use of the nomenclature A, B, C is that they do not mean the same things in every soil. For example, in arid environments, where evaporation and capillary movement of soil solution occur, the A, B, C horizons as conceived and defined become meaningless. In the humid tropics, as will be elaborated later, the system is inapplicable to the deeply weathered profiles. Thus for purposes of comparability some pedologists have suggested three main groups of horizons: horizons of accumulation (A), eluviation (E), and illuviation (I). Additional small letters are used to indicate the materials accumulated, eluviated, or illuviated: e.g. h stands for humus, s for soluble salts, c for carbonates, etc. Some soil horizons are more well-formed than others; weakly developed horizons are usually indicated by the use of brackets: e.g. (A) is a weakly developed A horizon. These ideas are partly incorporated into the A, B, C system as will have been noticed in Fig. 2.6.

Measurement of Soil Depth

Soil depth is not easy to define and many definitions and viewpoints exist. The main problem is what and what not to regard as soil. Is the parent material from which the soil is formed part of the soil profile or not? What portion of the rotten rock (weathered mantle) constitutes the soil parent material? Is a deep weathering (duricrust, laterite) profile also a soil profile? etc. Many more questions may be asked to which there are no precise answers.

In a large part of the earth's surface, the problem is not great. In most parts of the cool temperate, semi-arid, and periglacial regions the depth of the soil and its parent material may be only a few centimetres, which can easily be reached by digging with simple implements or investigated by the use of a hand auger. In the humid tropics, however, including areas which once had this type of climate, the weathered mantle may go down to hundreds of metres. Many of these profiles, called deep-weathering profiles, show evidence of having resulted from a single process, similar to the pedogenic process. To study these profiles, bigger equipment than the hand auger is needed. Given the difficulty of studying such deep profiles, it has been the practice to confine investigations to the topmost portion, especially that which contains living matter and supports plant growth.

Chapter 3

SOIL FORMATION I: ROCK WEATHERING

Soil formation generally involves two stages or phases. The first stage involves the accumulation of the raw (parent) material for the soil; while the second involves the formation of the soil from the parent material. The raw or parent material may result from *in situ* disintegration and decomposition of rocks and minerals, which process is generally described as rock weathering; or from the transportation and deposition of rock debris by the various agents of denudation—water, ice, and wind. This second source of soil raw material also involves the process of rock weathering, i.e. in the preparation of the loose rock particles for removal. In other words, knowledge of rock weathering is essential for the proper understanding of soil formation. We have therefore decided to describe it in this chapter, while soil formation proper is treated in the next chapter. In theory, it is possible to isolate the two processes or phases; but in fact they are inseparable. We are treating them separately here for the sake of simplicity, and also to afford tolerably detailed consideration of each.

The Process of Rock Weathering

The term 'weathering' has been defined in many ways. Two main viewpoints may be used as illustration. First, there is the viewpoint of the soil scientists, especially pedologists, who see weathering, as noted above, as the first phase of a two-stage process of soil formation. To these people, therefore, weathering is equivalent to 'the change of rocks from the massive to the clastic state' (Polynov, 1937, p. 12). Furthermore, weathering is seen not only as a destructive process, but also as a constructive one in which new, secondary minerals are synthesized from the products of the break-down of primary rock minerals. The other viewpoint may be taken as that of the other earth scientists, especially geologists, soil chemists, and soil physicists, who see weathering in terms of chemical, physical, physicochemical, and biochemical reactions, which may not necessarily result in the disintegration of the original rock. Thus Reiche (1950) defines weathering as 'the response of materials which were in equilibrium within the lithosphere to conditions at or near its contact with the atmosphere, the hydrosphere, and perhaps still more importantly the biosphere.' Ollier (1969) recently modified this definition, to the effect that 'weathering is the breakdown and alteration of materials near the earth's surface to products that are more in equilibrium with the newly imposed physicochemical conditions' (p. 1). Of these two viewpoints, the first is more relevant to our present discussion; so emphasis will be on those processes which produce clastic rock or loose rock debris and those which produce secondary minerals. These processes operate in what is generally referred to as the weathering crust.

The Weathering Crust

This term emanated from the Russians who defined it as the part of the earth's crust which is dominated by katamorphic processes. The zone has also been divided into two: an upper belt or zone of 'weathering' extending from the surface to the groundwater level, and a lower zone of 'cementation' from the groundwater level to the 'metamorphic' zone. The actual zone of weathering which is of interest here extends from the surface to the depth accessible to atmospheric oxygen. In other words, the weathering crust consists of the entire range of superficial deposits including weathered rock *in situ* and transported sediments.

A related term to the weathering crust is 'supergenesis' which embraces *in situ* weathering, pedogenesis (soil formation), sedimentation, and other phenomena related to the migration, transportation, and accumulation of chemical and biochemical elements and their compounds. Another related but not identical term is 'deep weathering'. While the weathering crust refers to the entire mantle of loose material overlying fresh unweathered massive rock, deep weathering refers specifically to a type of rock weathering which takes place under a given combination of environmental conditions, especially humid tropical conditions on low-relief surfaces. The concept of deep weathering has been discussed elsewhere (Faniran, 1971, 1974) and so will not be elaborated further here. Suffice it to note that it has significant implications for both the formation and the characteristics of soils.

Types of Weathering

The most common practice in the discussion of rock-weathering processes is to group them into physical and chemical weathering types. This means combining physical and mechanical weathering, and also chemical and biological weathering (Fig. 3.1). This practice is adopted here, although it is realized that this, like most classifications of natural processes and phenomena, is arbitrary.

The type of weathering in a place decides the composition, structure, and thickness of the weathered zone. The various factors of rock weathering discussed later in this chapter do not play equal roles everywhere, so that weathering differs from place to place. Lukashev (1970, p. 22) puts it thus:

The composition, structure and thickness of the weathering crust

SOIL FORMATION I

vary according to the parent rocks, the relief of the location of the parent rocks, the distance between the sites of deposition and the formation of the sediments, the mode of disintegration of rocks, the time elapsing between their formation and deposition, and the climatic and other conditions.

```
                          WEATHERING
                              |
            ┌─────────────────┴─────────────────┐
   PHYSICAL WEATHERING              CHEMICAL WEATHERING
            |                                   |
      ┌─────┴─────┐                    ┌────────┴────────┐
  PHYSICAL    PHYSICAL              CHEMICAL         BIOLOGICAL
 (THERMAL;  (MECHANICAL
  INTERNAL   EXTERNAL
  STRESS)    STRESS)
                                          |
   ┌──────┬──────────┬──────────┬─────────┬──────────┐
SOLUTION OXIDATION HYDRATION CARBONATION HYDROLYSIS REDUCTION
```

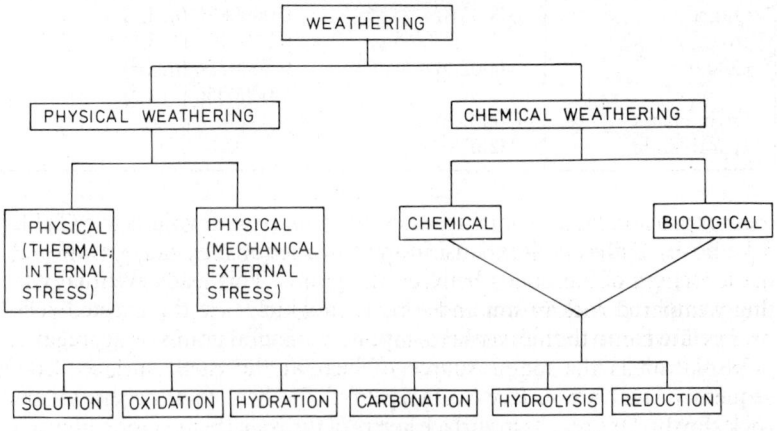

Fig. 3.1   Types of weathering

Physical Weathering Processes

These processes merely break rocks down to smaller fragments without changing their composition. Thus, if we chemically analyse two pieces of the same rock that have resulted from physical weathering the results should be compatible, i.e. the same elements and compounds should be present in compatible, if not the same, proportions, depending upon the original composition of the rock in question.

It is possible to distinguish two types of physical weathering, based on the operative process. First, disintegration may result from movement(s) from within the rock, e.g. expansion and contraction of constituent mineral grains or of the interstitial air in the rock. The high pressure produced by the interaction of the electrical double layers at the rock–water interfaces has also been suggested as a possible cause of this type of weathering, generally referred to as thermal weathering. In the other case, disintegration may result from the action of external agents, and so is referred to as mechanical weathering.

Thermal weathering, as the term implies, refers to rock break-down caused by temperature fluctuations due to differential heating and cooling of the rocks, in respect either of time or of space. Two causes may be distinguished. First, the mineral grains composing a rock expand and contract at different rates and in different directions, depending upon

TABLE 3.1
Expansion Factors of Some Common Rock Minerals

Mineral	Expansion Factors	
	Volumetric	Linear
Quartz	0·000310	0·0000070 (in L^3)
		0·0000137 (L^5)
Calcite	0·000200	0·0000256 (in L^3)
		0·0000550 (L^3)
Orthoclases	0·000170	
Hornblende	0·000284	

their expansion factor characteristics, both linear and volumetric (Table 3.1, above). Differential expansion and contraction may cause strain and the loosening of the bonds between the grains. This leads eventually to the granular disintegration of the rock, in which case the affected rock breaks down into the individual component mineral grains or aggregates.

Insolation is the major source of heat at the earth surface. Consequently, the rock surface heats up first; the heat then penetrates the rock slowly. This results in surface layers of the rocks being more intensely heated than the deeper parts, thus causing differential heating conditions as above. In addition, temperature fluctuations cause further disparities in the rates of expansion and contraction, which result in stress and, eventually, in the peeling-off of the top layers of the rock. This process is generally referred to as exfoliation or onion weathering. A similar process may result from the release of an overburden, e.g. the stripping of the weathered mantle cover of corestones and 'incipient' inselbergs. In both of these cases, the rock fragments may themselves be further broken down into smaller fragments by the process of granular disintegration.

The views discussed above are the traditional, old-established ones. At a recent symposium on geomorphic processes in arid lands, attention was drawn to additional possibilities.[1] One is the neglected role of the interstitial air in rocks. The expansion coefficient of air is greater than that of most rock minerals so that the expansion/contraction process should be more marked in the rock air, and so could contribute to rock disintegration.

Another factor is water. Not only is this capable of inducing chemical weathering and so assisting rock break-down, it has also been shown that interstitial water is capable of exerting tremendous pressure due to its electrical double layers. Such pressures, combined with those by thermal stresses both of the rock minerals and of the interstitial air, are thought

[1] See Schick, A. P., Yaalon, D. H., and Yair, A. (eds) 1974, *Zeit. für Geomorph.*, Suppl. Bands 20 and 21, Gebrüder Borntraeger, Berlin.

to be sufficient to explain thermal weathering particularly in arid lands.

Mechanical weathering, as noted above, results directly from the action of an external force or agent. Examples are:

(a) The use of explosives, hammer, or grinding machine to break rocks down into smaller fragments. Here it is *man* that is the agent of physical weathering. This type of rock break-down has increased in proportion in recent years, with the accelerated rate of road building and other civil engineering constructions all over the globe.
(b) When a tree root penetrates and grows into cracks and joints of rocks, expands them and eventually breaks the rocks down into small fragments. Examples abound of trees planted as 'flowers' or for decoration purposes which have caused great damage to concrete and walls of buildings when their roots got quite huge and powerful. Animals of various types also break rocks down but this is fairly insignificant.
(c) Ice exerts pressure on rocks to break them down. This process is commonly referred to as frost shattering, and it operates mainly in regions of high latitudes and altitudes, where temperatures drop below freezing during some weeks or months of the year. The main cause of the pressure exerted on rock walls in these regions is the increase which occurs in the volume of water when it turns into ice. The amount of increase is about 10 to 11 per cent, so that when water, which penetrates rock fissures, cracks, joints, and pores after a rainfall and gets trapped there, turns into ice, it has to make room for the extra volume. This it does by pressing on the rock walls. The pressure exerted may be so enormous and prolonged that it will cause rocks to disintegrate physically.
(d) Crystal growth, also in rock joints, cracks and pores. One example of this is salt crystal in arid environments. Evaporation during the day may cause salt to be precipitated in rock cracks, etc. As the salt crystals grow in size, they exert pressure on the rock walls and may eventually break them down.

The process of salt weathering is one of the aspects of physical weathering currently being studied in laboratories all over the world. One such study was reported recently by Andrew Goudie at the symposium referred to earlier. Among other things, it was discovered that salt crystallization and thermal expansion of salt, particularly sodium sulphate, are the most effective processes capable of disintegrating large boulders of suitable rocks, e.g. limestone, sandstone, chalk. Examples of such weathering have been reported from parts of Ethiopia, where lakes provide the salt which is actively engaged in the physical as well as chemical weathering of large boulders of lime-rich volcanic rocks.

In colloidal plucking, colloidal substances, by exerting tension on the walls of neighbouring rocks, loosen and remove mechanically rock flakes or mineral matter with which they are in close contact. In the past, it was believed that neither crystal growth nor colloidal plucking was as important as, say, frost shattering, which was taken to be perhaps the most important process of physical weathering.

The precise extent of thermal weathering has been in dispute, especially since the experiments in the 1930s by Blackwelder (1933) and Griggs (1936). In essence, the experiments involved subjecting rocks to great temperature ranges, the types rarely encountered on the earth surface. It was reported that in the absence of water, the resulting expansion and contraction in rocks did not produce stress great enough to cause shear or cracks in the rocks. It was therefore concluded that water and chemical reactions were essential concomitants of physical weathering. This view has been supported by field and laboratory observations, including those by Ollier (1963) in Australia. There is currently an upsurge of interest in physical weathering. Additional causes of rock failure are being found. But it is becoming more and more difficult to draw a line between physical and chemical weathering, both of which seem to operate *pari passu*.

Chemical Weathering Processes
The following initial statements about chemical weathering are useful. (1) Unlike physical weathering, chemical weathering affects the chemical and mineralogical composition of the original (parent) rocks, according to the type and the degree or intensity of the weathering action or process. Some minerals are completely dissolved; some are changed chemically; while others still are formed from the resultant chemical solutions and products. (2) The main agents of chemical weathering are water, gas, and organic acids. Depending upon which of these are involved and also their amounts, chemical reaction may be slight or intense. Chemical weathering processes are classified partly according to the agent involved and partly according to the intensity of the process, and the end-products. (3) Physical weathering aids chemical weathering, just as chemical reactions participate in the physical break-down of rocks. Rocks that are more heavily fractured and/or fragmented by physical processes are more likely to be intensely acted upon by chemical processes, given suitable environments. This is another way of saying that the greater the effective surface area of the rock, the more intense is the chemical weathering process likely to be. The effect of grain size and other attributes of the parent rock on weathering is further discussed below. (4) Finally, many types of chemical weathering are known, the most commonly described ones being hydration, oxidation, carbonation, hydrolysis, solution or dissolution, and reduction. These are now described briefly.

HYDRATION

This is perhaps the simplest chemical weathering process; it occurs when water combines with other constituents which, in the present case, are the elements and compounds making up the rock. As water is central to all chemical weathering, hydration is a dominant process in the humid regions of the world, leading to the formation of hydrous compounds, e.g. kaolin and serpentine, gibbsite or hydrated aluminium oxide, and goethite or hydrated iron oxide.

Hydration is a mild form of chemical weathering. Although capable of weakening or even disintegrating rocks to great depths, it may not affect the chemical composition of the original rock, apart from the absorbed water. Merrill (1897, p. 188) described how granitic rocks have become disintegrated to a depth of many centimetres with loss only of comparatively small quantities of their original chemical constituents and with apparently little change in their form of combination. Hydration causes increase in the volume of the rock, in the order of up to 150 per cent. In most cases, however, volume increase is less than 50 per cent.

Although hydration is a relatively easy chemical process, the opposite or reverse process of dehydration is much more difficult especially under normal weathering conditions. For instance, for goethite $(Fe(OH)_3)$ to be dehydrated to haematite (Fe_2O_3) under laboratory conditions, it is necessary to dry the goethite at very high temperatures—of the order of 400°C (700°F). Dehydration is therefore possible only in areas of very high temperatures and possibly little or moderate rainfall.

OXIDATION

This is another relatively simple process in which both oxygen and water take part. Of the two substances, oxygen is by far the most important oxidizing agent in nature. Oxygen forms, by volume, about 21 per cent of the air near the earth's surface; it is also present in water. When air is dissolved in water, as in rainwater, the free oxygen content of the water increases, in some cases to between 30 and 35 per cent. It is the free oxygen in such waters, and also in the air, that acts as the oxidizing agent.

The intensity of oxidation in rocks is also related to the type of rock upon which it acts. Oxidation is generally most intense in rocks whose constituent elements possess different valencies or combining powers. Oxidation is generally most pronounced in rocks containing the original or native elements, e.g. native iron (Fe). Such minerals therefore undergo oxidation more readily than the other common rock constituents. Oxidation operates best in well-drained conditions as in the area above the zone of permanent or even periodic water saturation. The diagnostic feature of oxidation is the red or brown colour of some soils and rocks at or near the ground surface.

There are many examples of the oxidation of rock minerals. The most commonly cited example is pyrite (FeS) being oxidized into ferric hydroxide ($Fe_2O_3nH_2O$). The actual process is as follows: $FeS_2 + nO_2 + mH_2O \rightarrow FeSO_4 \rightarrow Fe_2(SO_4)_3 \rightarrow Fe_2O_3nH_2O$, which in simple language states that pyrite in the presence of water and oxygen oxidizes first into ferrous sulphate, then into ferric sulphate, and finally into ferric hydroxide. Other iron minerals which undergo oxidation are marcasite and pyrrhotite. In all these cases, the last phase is the stable state in which the mineral can exist in the newly imposed physicochemical environment.

The end-products of oxidation are mostly stable elements and compounds, in which form they exist for quite a long time. The indurated zone of a deep-weathering profile consists essentially of oxidized substances, e.g. haematite, goethite, gibbsite, manganese oxide, and titanium oxide, among others, some of which have been and are still being exploited as ores, wherever they occur in economic quantities (cf. Faniran, 1970).

CARBONATION

This is the process by which carbonate (CO_3^{--}) or bicarbonate (HCO_3^{--}) ions combine with a rock element usually to dissolve it. The main factor here is carbon dioxide (CO_2), dissolved in rainwater. The carbonic acid thus formed is a weak acid, since CO_2 forms only 0·03 per cent of air, and 0·45 per cent of rainwater, by volume. The solvent action of carbonic acid is enhanced in vegetated areas, where organic acids are also involved. Under this condition, desilicification and hydration also take place. Carbonation is particularly important in limestone regions, where it is largely responsible for the extensive weathering (dissolution) that occurs.

HYDROLYSIS

The process of hydrolysis requires the presence of water. It occurs as a result of the reaction between the dissociated hydrogen (H^+) and hydroxyl (OH^{--}) ions of water and the mineral elements of a rock. Not all rocks react to water this way, but for the feldspars and other rock-forming silicates, water is generally recognized as 'public enemy no. 1' (Keller, 1957).

The bulk of the primary rock-forming minerals are of course the silicates, so that hydrolysis is a very important process in the zone of weathering. However, hydrolysis does not necessarily take place when these rocks come into contact with water, as shown by fresh rock in water-saturated zones. The drainage situation is very important. For instance, there must be repeated supply of fresh water to leach away the soluble products of the hydrolysis reactions. This situation is possible when there

SOIL FORMATION I

are the following:

(a) repeated leaching by fresh rain and/or snow-melt water;
(b) introduction of H ions which will combine with OH ions, thereby removing them as water;
(c) precipitation of ions as relatively insoluble compounds;
(d) removal of ions by chelates;
(e) absorption and assimilation of the products by living organisms, or absorption of the products by colloidal substances.

In short the best environment for hydrolysis is one where there is free movement of underground water.

The process of hydrolysis usually involves one of the most intense chemical reactions in the zone of weathering, as a result of which some minerals are decomposed and others are changed so much that they bear little or no resemblance to their original forms. Generally, the process involves a chain of chemical reactions, similar to those described above for hydration and oxidation. The reactions continue until stable elements are formed, e.g. where feldspars in the presence of water decompose into kaolin and opal. It is important here to note the similarity between hydration and hydrolysis. Indeed, both these processes are sometimes considered together in which case hydrolysis is taken as a more intense and more advanced form of hydration.

DISSOLUTION OR SOLUTION

Most substances in rocks are soluble, although the degree of solubility differs widely. The most susceptible substances are chlorides and sulphates, while the least are the silicates, e.g. quartz. If we take common salt (NaCl) as an example, what happens on its coming in contact with water (H_2O) is that the positively charged hydrogen ions (H^+) combine with the negatively charged chloride (Cl^{--}), while the negatively charged hydroxyl (OH^{--}) combines with the positively charged sodium (Na^+) ions to effect neutralization (solution) of the elements.

The process of solution is important for a number of reasons. First, because highly soluble substances are rapidly removed, the chemical situation in the soil is altered. Second, dissolved substances tend to ionize and so can react or combine with other elements, thereby activating the weathering reactions. Solution environments are therefore rarely in static equilibrium chemically. Third, many solutions occur only under restricted environmental conditions, e.g. iron and manganese under reducing conditions. Finally the process of solution or dissolution is a part of, as well as an important condition for, the other chemical processes, e.g. carbonation and hydrolysis.

REDUCTION
This is important because it operates both in the zone of weathering and in the actual process of soil formation. Unlike oxidation in which oxygen is absorbed by a substance from which hydrogen is removed, reduction is the reverse of this process, i.e. the removal of oxygen from a substance and the introduction or addition of hydrogen to it. Oxidation and reduction are therefore sometimes considered together, especially in relation to the oxidation–reduction potential or Eh.

In the zone of weathering, reduction is facilitated by anaerobic or completely saturated conditions. For instance, under a reducing situation, ferric hydrate can be reduced to elemental iron, part of which may occur in colloidal form. Sulphates are also usually reduced, leading to the release of hydrogen sulphide gas. Reduction involves the action of living organisms, e.g. bacteria, in which case it may be considered as an example of biological weathering, or more precisely soil formation. Reduction is, however, essentially a weathering process; the overlap is merely indicative of the close bond between rock weathering and soil formation.

Factors Influencing Rock Weathering

The factors that affect or influence the rock-weathering process can generally be grouped into two categories: (1) those pertaining to the geological–geomorphological environment, e.g. the nature of rock and minerals and the local topography; and (2) those relating to the energy sources, e.g. climate and plant and animal life. The time factor conditions the operation of the process as well as the age of the parent rock. It therefore appears to belong to both categories.

Rocks are made up of mineral elements and compounds. The nature and properties of all these are important in their weathering. The following are a few of the characteristics of rocks and minerals which significantly influence the course of weathering:

(a) the packing of the constituent atoms of elements and compounds;
(b) the availability or otherwise of void or open spaces in the rock or mineral aggregates;
(c) the chemical composition and chemical structure of the minerals;
(d) the size, shape, perfection, and arrangement of the constituent crystals and grains;
(e) the presence of some readily weathered minerals in the structure.

We shall not discuss each in detail, but only note the following general observations about them.

(1) Rocks in which the constituent minerals are well packed, so as to leave little or no pore spaces, are generally resistant to weathering; while

rocks with loosely packed grains are readily penetrated by water and gases which effect their weathering.

(2) Minerals with similar chemical composition but different crystal structures have varying degrees of solubility. For instance, calcite and aragonite have the same chemical composition—$CaCO_3$—but because the atoms are arranged differently in the two minerals, aragonite is about ten times more soluble than calcite.

(3) Large minerals are generally harder to weather than small ones. This is because weathering is a surface activity and minerals with small crystals have a much greater surface area than those which, although having the same volume, consist of large crystals. As a general rule, if a grain of mineral 1 mm across is broken into particles of 0·1 mm across, the surface area increases at least 1,000 times, and so the intensity of weathering. However, when a certain fineness is reached, e.g. clay-size particles (<0·002 mm), the particles tend to become generally inert to further changes, at least under normal weathering environments. Similarly, platy crystals have larger effective surface areas than chunky ones and so are more readily weathered.

(4) Where silicon (Si) forms the central atom of a compound or mineral, the unit is generally stronger than when aluminium (Al) or iron (Fe)—both readily oxidized elements—forms the central atom.

(5) Finally, rocks with weak structures or empty (void) spaces in the structure are susceptible to weathering.

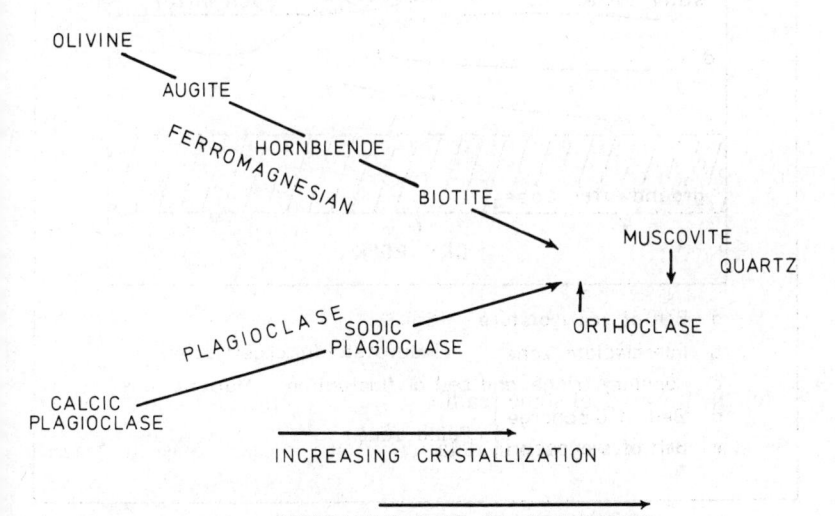

Fig. 3.2 Weathering sequence of common rock types

Thus, if a rock consists essentially of hard minerals such as quartz or muscovite, such a rock will be resistant to weathering—e.g. quartzite, granite, etc. Conversely, if a rock is made up of soft minerals such as olivine it will be readily weathered, e.g. basalt. The weathering sequence of the common rocks and minerals is shown in Fig. 3.2. More often than not, however, minerals of varying degrees of weatherability occur together in a rock. This means that at any point in the process of weathering these minerals will be altered to varying degrees, as has been shown among others by Goldich (1938) and Reiche (1950, pp. 46–53).

The effect of topography on weathering is less direct than that of rock type. First, it conditions the position of groundwater, which then influences the processes operating in the zone of weathering. The relationship

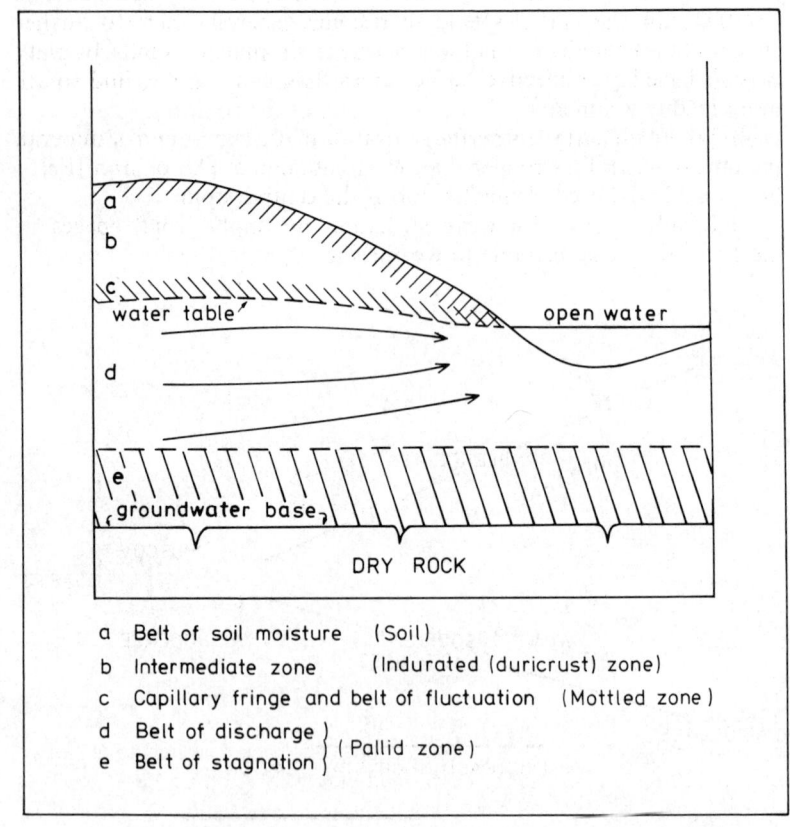

a	Belt of soil moisture	(Soil)
b	Intermediate zone	(Indurated (duricrust) zone)
c	Capillary fringe and belt of fluctuation	(Mottled zone)
d	Belt of discharge	(Pallid zone)
e	Belt of stagnation	

Fig. 3.3 Topography, groundwater, and weathering zones (modified from Ollier, 1969)

between topography, groundwater, and weathering zones is illustrated in Fig. 3.3. Second, topography controls the rate of removal of the weathered products. Generally speaking, removal is faster on steep than on gentle slopes. This is therefore an important factor in deep weathering since it is unlikely that we have thick accumulations of weathered rock or saprolite, and consequently soils in areas of high relief and steep slope.

Energy and Rock Weathering

With regard to rock weathering, the main sources of energy are atmospheric gases, especially oxygen and carbon dioxide, water, especially hydrogen (H^+) and hydroxyl (OH^-) ions, and the various organic acids. These have been discussed in relation to the different types of weathering above, especially how variations in the combination of these factors result in different weathering processes and products.

However, the activities of these factors are strongly controlled by what may be generally described as climatic elements—e.g. temperature, humidity, or moisture, and the pH and Eh of the solutions involved in chemical reactions. The following observations have been made on the role of climate on rock weathering:

1. The efficiency of either physical weathering (disintegration) or chemical weathering (decomposition) is to a large measure determined by the climate.
2. The rate of chemical reactions tends to increase as temperature increases, at least up to a point. As a general rule, chemical reactions double or even treble if temperature increases by $10°C$.
3. Biological activity also increases with temperature, in a fashion similar to (2) above.
4. The supply of organic acids depends on climate as a controlling factor of vegetation and animal life. The greater the quantity of organic acids made available to the zone of weathering, the more intense the weathering.

Consequently the depth and structure of weathering varies from place to place, being greatest in the humid tropics and shallowest in the extremely cold and/or dry regions.

The Product of Rock Weathering

The general descriptive term for the product of rock weathering is saprolite, including the substances removed either in solution or in colloid suspension, and residual material. However, some people restrict saprolite specifically to the latter (residual) material, which, as the parent material for soil, is also of particular interest to us here.

Because of the complexity of weathering types and intensities, the weathered mantle or saprolite is usually a mixture of rock and minerals in all phases and stages of decomposition. Nevertheless, the most important materials are mostly stable minerals, including those that are too hard to be altered under normal atmospheric conditions—e.g. quartz, zircon, rutile (or primary minerals), or the clay minerals and the hydrated sesquioxides. The latter are mostly secondary minerals. We shall describe some of these components briefly.

The clay minerals are illites (also known as hydrous micas and bravaisites), montmorillonites, and kaolinites. They have been considered in Chapter 2 and so need not be discussed further. The sesquioxides of iron, aluminium, and manganese are widely distributed: titanium oxide is another product of weathering but is generally rarer than the first group. They are all formed in free-drainage (oxidizing) situations and from rocks which contain the native forms of the minerals. Aluminium hydroxide or gibbsite appears to form directly from the weathering of basic igneous and metamorphic rocks, or indirectly from the break-down of kaolinitic clays. The formation of these oxides and hydroxides, especially when associated with typical deep-weathering profiles, has been attributed to tropical or at least subtropical climates, especially where a number of other favourable factors of weathering are present including: weatherable rock; extensive low topographic relief such as is associated with erosion surfaces and other plains; fairly open vegetation; and a fairly long period when these conditions are stable and unchanged.

Also associated with the end-products of weathering are new substances which result from the coagulation of colloidal silica, alumina, and ferric iron oxide, and from the reaction of these with dissolved materials, e.g. phosphoric acid. There are also deposits from indigenous solutions, e.g. the amorphous silica of some 'desert hardpans' and the calcium carbonate (calcrete) of the lower horizon of some weathered profiles and soils. The latest approach is to relate at least some of these hardpans or duricrusts to a previous period of deep weathering, especially where they are underlain by typically mottled zones and pallid zones, each of which may be anything from one centimetre to more than 100 metres in depth.

Finally, it should be noted that the end-product of weathering is the same thing as the mineral matter part of the soil. It is not soil until it is further worked upon by pedogenic processes. The processes and factors of soil formation are discussed in the next chapter.

Chapter 4

SOIL FORMATION II: SOIL-FORMING FACTORS AND PROCESSES

It cannot be overemphasized that there may not be any marked break between the process of rock weathering as described in the last chapter, and that of soil formation. The division is made here partly because weathered rock is not soil but has to be turned to soil by the operation of certain factors and processes. There is also the fact that the parent material of soils is not necessarily weathered rock *in situ*; sediment deposited by rivers, wind, ice, and gravity (e.g. landslide, rockslide, mudflow, etc.) as well as beach sand, shingle and other deposits are other possible sources of soil mineral matter. It is on these materials that energy-releasing elements of climate and living organisms act to form the soil body. The division may also be justified on the grounds that rock weathering should have operated on the earth's surface, perhaps for many millions of years prior to the time life appeared on the earth. During that period, it is difficult to imagine soil being formed, since its formation requires, among other things, the organic component of the earth–atmosphere system or geosystem.

However, in the evolution of soils other factors besides organic matter or the biosphere are also important. Indeed, as noted long ago by Dokuchaev, 'the soil is the result of the combined activity and reciprocal influence of parent material, plant and animal organisms, climate, age of the land, and topography', which collectively influence the operation of pedogenic processes such as podzolization, calcification, ferralitization, salinization, etc.

It is rather irrelevant to consider which of these factors was first, and in which order they became active since the dawn of life on our planet. But if it is at all possible to reconstruct the history of soil formation at the earth's surface, the first factor is likely to be the parent material resulting from millions, perhaps billions, of years of physical and chemical weathering. This material could have existed either as *in situ* weathered mantle or as deposited sediments, both derived either directly or indirectly under the influence of weather or climate. In whichever form it existed, it should have provided a more suitable abode for plant life, especially the higher plants, when they finally arrived on the scene. The advent of plants then set off biochemical reactions which, with the normal geochemical reactions of rock weathering, led to the formation of the soil body.

However, ever since the formation of the first soil, it is difficult to imagine a succession of geochemical and biochemical reactions. Rather, except in the most extreme situations where life is precluded, these two processes would be expected to act contemporaneously in many places, although one or the other of them may predominate at a given time and stage in the development of the soils. Geochemical reactions will predominate over biochemical reactions during the early phase of the process, and biochemical reactions during the latter phases of soil formation. Biochemical reactions are very important in soil formation, as evidenced by the fact that where they are absent or negligible, e.g. in very dry deserts and ice-capped regions, soils hardly exist. This means that whereas rock weathering exists in one form or another all over the globe, soil formation does not. Finally, the two processes may be distinguished according to the transformation which attends them and their end-products. Whereas rock weathering 'simplifies' mineral complexes to produce simple compounds, pedogenesis or soil formation is synthetic and so produces complex structures, compounds, and materials.

The Factors of Soil Formation

The factors which Dokuchaev listed at the turn of the century (see Chapter 1) are still being considered with respect to soil formation today. These factors, which are also the same as those considered in connection with rock weathering, have been grouped into soil-forming or active (energy-supplying) factors and relatively passive (energy-receiving) factors. The division here, like any other similar division, is mainly for convenience. Although climate and organisms are definitely important in the process of soil formation, the other factors—parent material, time, and topography—do exert strong, and in some cases decisive, control on process to make their influence felt in the resulting soils. All the factors are important in their own right; they all mutually interact to form the soil body.

Since these five factors have been identified, many workers have studied them and expatiated on them, especially in terms of the soil–soil formers equation. The original Dokuchaev equation was

$$s = f(pm, c, b, a, t) \quad \ldots \ldots (4.1),$$

where s = soil; f = function, pm = parent material, c = climate, a = age of land, b = biosphere, and t = topography. This was later modified by Jenny (1941) as

$$s = f'(cl'', o, r, p, t) \quad \ldots \ldots (4.2)$$

where the comma denotes internal function, cl'' is soil climate, o is soil organism, r is the shape of the soil surface, and p and t stand for parent material and time respectively. Jenny also proposed a different (external environment) version of the latter equation as follows:

$$s = f(cl, o, r, p, t) \qquad \ldots \ldots \quad (4.3)$$

in which cl is external climate, o stands for all organisms (possibly man inclusive), r for relief forms rather than the slope, and the rest are as in his first equation. He also recognized the possibility of discovering other factors and so left the equation open. Accordingly, some people have advocated the inclusion of man as a separate factor from other organisms. Other factors suggested for separate inclusion are space or the locational factor, gravity flow and other forms of mass wasting, soil climate, and underground hydrology. Crocker (1952) has for instance proposed that it is more useful to express soil as an integral of five factors against time. He proposed the equation

$$s = f(c, o, r, w, p)t,$$

where f is integral function, c is climate, w is water table, and the rest are again as in Jenny's equation.

The relative importance of these factors is as problematic as their sequence in time. The factors appear to be arranged in a certain fashion or according to some ordering system in the equations proposed, especially those based on Jenny's approach. For instance, climate and organisms are listed first and parent material and time last, suggesting a decreasing order of importance. Furthermore, as shown in Chapter 1, emphasis has shifted from one factor of soil formation to another. Prior to Dokuchaev, emphasis was mainly on parent material which was then number one factor. Dokuchaev and most later workers stressed the importance of both climate and biological activity. Some workers have also emphasized time, although it is listed last in all the equations. It may therefore be that the order in which the factors appear in the equations reflect other things besides degree of importance, e.g. the complexity of the factors and the extent of their influence on the other factors. Climate is definitely more complex in its effect than, say, topography. In any case the order of importance of these factors is not very crucial here, although it is worth thinking about.

One other difficulty that may be mentioned is the usefulness of the equations proposed, especially when they cannot be operationalized. However, we are now able to relate quantitatively certain soil properties to certain environmental properties, e.g. clay content and rainfall, organic matter content and temperature, soil depth and degree of slope, etc. We can also confidently express soil productivity in terms of certain internal and external variables of soil formation (see Chapter 9). The recognition of these factors and their detailed study have greatly advanced the course of soil science. There is therefore no doubt that if we stop taking the equations too literally, there is a lot to gain from their application. The various components of the equations are described below under two major headings of 'active' (energy-supplying) and 'passive' (energy-receiving) factors.

The Active Factors
These are the factors which activate the other factors of soil formation to create the 'living' soil body from the 'dead' rock debris and minerals commonly referred to as parent material. The activation involves energy which is supplied in many forms and from many sources. The treatment of these factors so far has been mainly on the basis of whether the energy supplied pertains to the atmosphere in which case it is discussed under the climatic factor; or whether the energy is supplied by organisms of the biosphere. In other words the emphasis is on the energy sources.

By contrast it is possible to think of the form or nature of the activating energy itself: e.g. heat energy and the role of solar radiation and temperature; chemical and organochemical energy and the role of water, gases, and acids, especially organic acids; and mechanical energy and the role of roots and animals, including man. The effect of gravity and capillarity, especially in relation to the downward and upward movement of chemical and organochemical solutions, may be treated as types of mechanical (kinetic) energy, in which case the movement involves either part of the soil body as in the case of soil mixing by soil animals, or soil solutions and mineral elements as in soil leaching and soil erosion. These translocations are by far the most important processes in the formation of the soil body, especially the differentiation of soil horizons. For the sake of simplicity, however, both the sources and the forms of energy involved in soil formation will be described.

CLIMATE, ENERGY, AND SOIL FORMATION
Of the various elements of climate usually included in the discussion of soil formation, precipitation (including soil water), humidity, and temperature are by far the most important. Wind is usually mentioned mainly in connection with the soil moisture component and also in connection with soil erosion. Wind is also mainly a soil destroyer and so need not be discussed here, soil erosion being the topic of a subsequent chapter in this book (Chapter 10). Only precipitation and temperature will therefore be considered here.

PRECIPITATION AND SOIL FORMATION
Moisture is one of the primary elements of climate that greatly influence the character of the soil body. The main source of soil water is precipitation, although thawing ice may be locally important, e.g. in temperate

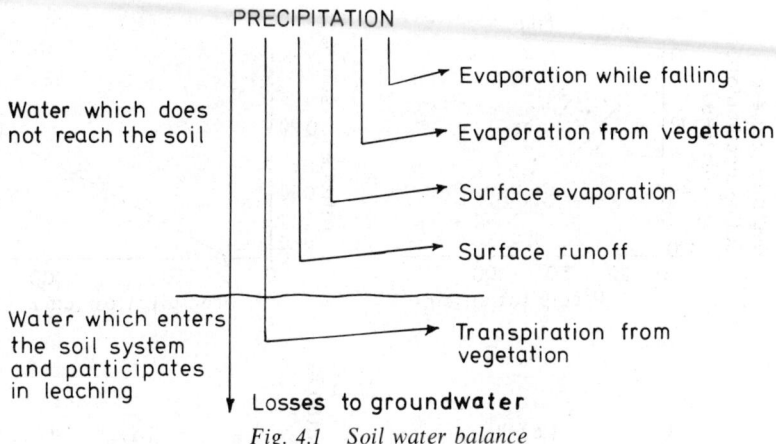

SOIL FORMATION II

```
                    PRECIPITATION
                    │ │   │ │
                    │ │   │ └──→ Evaporation while falling
Water which does    │ │   │
not reach the soil  │ │   └────→ Evaporation from vegetation
                    │ │
                    │ └────────→ Surface evaporation
                    │
                    │ ┌────────→ Surface runoff
                    │ │
────────────────────┼─┼──────────────────────────────
                    │ │
Water which enters  │ │
the soil system     │ └────────→ Transpiration from
and participates    │             vegetation
in leaching         ↓
                    Losses to groundwater
```

Fig. 4.1   *Soil water balance*

latitudes. But not all the water made available by these sources participates in soil formation. Losses are incurred in many ways as illustrated in Fig. 4.1. The study of water balance and the water budget is an important branch of hydrology and climatology; studies so far made show that not only does the total amount of available water vary widely from place to place but also the amount that is available for soil formation varies, governed by such factors as the infiltration and through-flow characteristics of the soil surface, the rate and amount of runoff, the amount of interception, and, of course, the amount of evaporation and evapotranspiration.

Despite the fact that soil water is not the equivalence of precipitation, studies made so far show that a number of processes and reactions associated with soil formation can be directly correlated with the amount of precipitation: e.g. (a) the depth to which calcium carbonate is leached, (b) the concentration of nitrogen, (c) the formation of clays, and (d) the accumulation and decomposition of organic matter, all of which are very important indices of soil formation. The overall effect of these processes is roughly sketched in Fig. 4.2. Global variations in soil processes and soil types are further discussed later in this book (Chapter 8).

Precipitation affects soils indirectly through its influence on rock weathering and the type of parent material available for soil formation, as well as through plant growth. Moisture is one of the controlling factors of plant growth, which in turn harnesses the solar energy that acts on the mineral matter.

Soil water acts as the carrier of substances, both in solution and in colloidal state, from one part of the soil profile to another, thus depriving one portion of the same constituents with which it enriches another part.

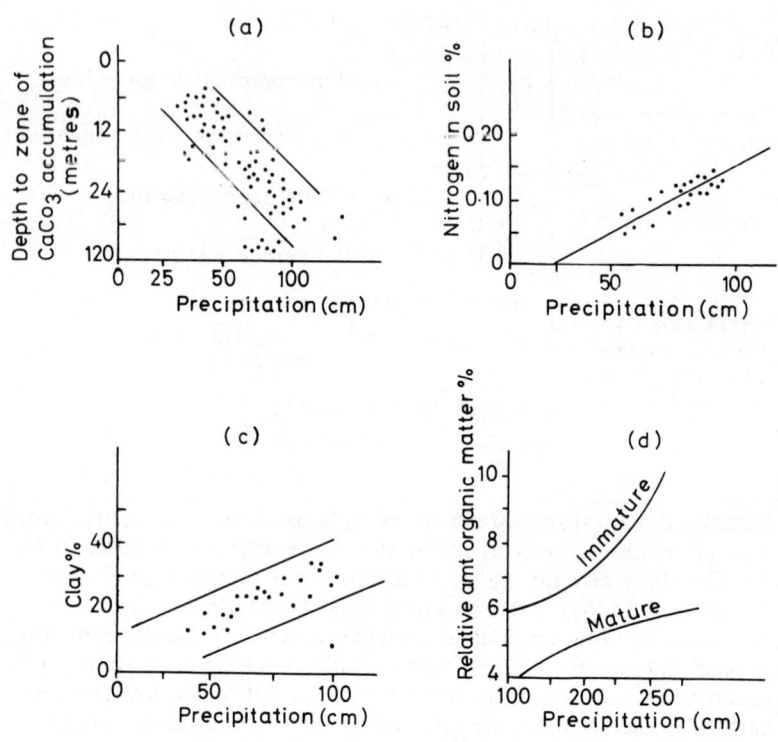

Fig. 4.2 Precipitation and soil formation

In a leaching situation, substances are leached from the top part of the soil body (the A horizon) and deposited in the lower part (the B horizon), or totally lost to drainage water. Depending upon the amount of water involved, and also the acidity of the solution, especially in the case of leaching, sufficiently large quantities of soil material may be finally lost to the subsoil, groundwaters, and the oceans as to make the soils of an area very infertile. Conversely, capillary translocation involves the removal of substances from the lower to the upper part of the soil profile. We may therefore see soil water in terms of: (a) its effect on the activation of chemical reactions leading to the break-down of rock and other solid fractions of the soil parent material, that is, in terms of inducing further weathering of soil parent material; (b) its role as a medium for the loss of certain soil elements; and (c) its significance in controlling the rate and nature of translocations which take place in the process of soil formation.

Of all the sources of water loss to the soil, evapotranspiration is among the best quantified. Thornthwaite's precipitation-effectiveness index is one such example. Although found to be not wholly applicable in certain areas, this index has been used to classify soils (see Bunting, 1965, p. 59). Table 4.1 shows the type of soil environments distinguished from this exercise. Note the arbitrary and irregular nature of the boundary marks used.

TABLE 4.1

Hydrological Soil Series

P/E ratio	Soil
less than 0·20	Extremely arid (A)
0·21–0·40	Arid (B)
0·41–0·75	Moderately arid (C)
0·76–1·20	Moderately moist (D)
1·21–1·95	Moist (E)
1·96–2·90	Very moist (F)
greater than 2·90	Especially moist (G)

Extracted and modified from Bunting, 1965, p. 65.

TEMPERATURE AND SOIL FORMATION

The effect of temperature on soil formation may be direct or indirect. Directly, temperature influences the rate both of chemical activity and of bacterial activity. Indirectly, temperature affects the rate of evaporation of soil water and consequently its leaching effect, causing in some cases an upward movement of some soluble salts which modify the effects of the other factors of soil formation. Furthermore, high temperatures encourage luxuriant growth of vegetation and so facilitate the abundant supply of organic matter to the soil.

An aspect of temperature/organic matter relations is the rate of organic matter decomposition. This rate increases with temperature, so that the tropical regions with luxuriant vegetation growth are also the regions with optimum conditions for decomposition of organic matter by bacteria and microbes. By contrast, the low temperatures of cold regions mean a low rate of bacterial and microbial activity and, consequently, the accumulation of organic matter, sometimes in the form of peat. This accumulation of organic matter has pronounced effects on the character of soil profiles developed in these regions.

The indirect effect of temperature may also be felt in the process of soil freezing. A function of both soil water and, perhaps more importantly, soil temperature, the main effect of soil freezing is soil heaving, causing the seasonal expansion and contraction of the soil body. The extent of this influence varies according to the composition of the soil. Generally, heaving affects soils rich in organic matter more than it affects inorganic materials. For instance, while heaving up to 50 mm/yr has been reported for some organic soils in northern Germany, figures for loess and sand are respectively 42 mm/yr and 23 mm/yr. Furthermore, extremely low temperatures in the subarctic regions of the world produce permafrost conditions, which limit the zone of active soil formation and so the depth and character of the resultant soil body.

THE BIOSPHERE AND SOIL FORMATION

A common approach to this topic in the literature has been to group the plants and animals which participate in soil formation, and to describe each of them according to the role played in soil formation. Although we accept the differences in the activities of these various groups (see Chapter 2), emphasis here will be on the general effects of all organisms on the process of soil formation.

Their role can be viewed, as in the case of the climatic elements, as direct and indirect. The direct role of plants in soil formation is mainly in the supply of organic matter. By far the most important source of soil organic matter is the plants of the earth's surface. The supply is mainly from leaf fall, dead roots, tree trunks, twigs, fruits, and seeds. These are acted upon by the bacterial and other microbial components of the soil and worked into the soil. Dead animals are similarly acted upon and worked into the soil. The process by which organic matter is worked into the soil is called humification, a process which involves also the activity of the fauna component of the soil. Bacteria and fungi (plants) break down the plant tissue upon and within the soil, to start with. Animals such as earthworms and termites then work the decomposed organic matter into the soil, as they dig up and mix soil horizons and materials. The decomposed plant materials also supply carbon, nitrogen, sulphur, and acid substances which are very important in the process of soil formation. Organic acid, for instance is very crucial to the process of leaching.

The indirect role of the biosphere in soil formation takes many forms. Tree roots penetrate rocks and soils and so facilitate the penetration of soil solutions and colloids which then extend the depth of the soil profile. Vegetation also affects the various elements of climate. Generally, forests tend to make climate milder; they tend to reduce wind speed and so lower the rate of evaporation, especially in warm regions and seasons; they cut down on the quantity of precipitation that reaches the soil surface through interception; and, perhaps most importantly, they protect the soil against erosion.

A change in the vegetation cover of a place has been observed to be accompanied by a change in soil type. The often-quoted examples are where the replacement of coniferous plantations by deciduous trees has produced podzolic soils from the original podzols and where oak woodlands, replacing a grass cover, have led to marked leaching of the original chernozems. The changes that occur as a result of forest clearing especially for crop cultivation can be even more marked. First, the supply of organic matter is interrupted; the plant nutrients in the soil are used up by crops, and may not be returned as in a natural forest. Instead, artificial fertilizers may be applied, thus changing the nutrient status of the soil. The role of man in soil formation and soil destruction is becoming an increasingly important aspect of soil study. Man, like the other large animals, mainly disturbs rather than aids the formation of the soil profile.

Conclusively, the role of organisms in soil formation is related to that of climate. In the first place, climate has a strong control over vegetation. It also controls the rate of organochemical reactions, as well as the number and type of fauna which in one form or another influence soil formation. In short, the role of organisms in soil formation is perhaps best illustrated by such chemical and organochemical processes as nitrification, denitrification, humification, and mineralization, in which they help in converting substances from one medium or form to another.

The 'Passive' Factors

These are traditionally taken to include the soil parent material, the topography, and the time element. The assumption here is that both topography and time affect the nature of the parent material rather than provide energy which changes the 'dead' rock debris to the 'living' soil. This assumption like others of its kind, has been noted to suffer somewhat from oversimplification. The parent material, it can be argued, is not devoid of energy. In fact rocks have some endothermal or endogenic sources of energy, derived from two main sources: gravitational, derived from the compaction of mass and the addition of angular momentum during the formation of a planet; and radiogenic or atomic energy, derived from the break-down of the atomic nuclei of certain earth materials, e.g. uranium, radium, and thorium. However, all these other sources make up just about 0·1 per cent of the energy involved in the processes operating at and near the earth surface, while solar energy supplies 99·9 per cent. Finally, if soil formation refers to the mechanical, chemical, and organochemical processes which change the inorganic (dead) parent material into soil with characteristic morphological and compositional features, the role of parent material, topography, and time can be seen as rather passive. This does not mean, however, that these other factors are any less important than the energy suppliers, as will be shown.

PARENT MATERIAL AND SOIL FORMATION

The soil parent material, as stated earlier, is either the product of *in situ* weathering, or transported and deposited material. Characteristically, it is unconsolidated (clastic) material found in a place where pedogenesis has not set in—e.g. newly emerged beach sand or river sand; a newly deposited glacial till; a fresh sand dune; or below the true soil horizons.

A warning is called for here, especially in the case of transported material. Many soils have developed from materials different from the country rock, e.g. a glacial drift or loess overlying unweathered, or even weathered, granite, sandstone, or any other type of country rock. However, the parent material of many soils is derived directly from the weathering of the country rock, in which case the nature of the original rock is reflected in the soil in one form or another.

One major reason for describing parent material as a passive factor of soil formation is perhaps the fact that on a world or continental scale different climatic and biological conditions tend to make different soils from the same type of parent material, while similar soils may form from different rock types, given the effects of similar climatic and organic influences. The best illustration of this is in zonal or climatogenetic soils described in Chapter 8. Parent material, however, needs not be as passive as this; its effect may still be strong on soils.

The best examples of soils that have been described as reflecting parent material are generally immature or young soils (or regosols and lithosols). Where, because of the slope factor, soil erosion is rapid, the soils tend to reflect the parent material rather than the pedogenic process. Furthermore, where there has not been sufficient time for soils to form after the dissection of an erosion surface, the soils may reflect the parent material.

A good example of this last situation is provided by the soils of the Sydney district of New South Wales, Australia. The area concerned was deeply weathered, perhaps sometime during the Tertiary, and subsequently dissected, so that the deep-weathering profiles are widely truncated and/or stripped (cf. Faniran, 1969). Consequently, the soils of the area at the present time tend generally to vary from those formed in indurated-zone materials on the hilltops, plateau-tops and interfluves, through those formed on mottled-zone materials in the intermediate slope regions, to those formed in pallid zones and ordinarily weathered rocks on the lower slopes, in a catenary sequence. In other words, the present-day soils in this area seem to reflect the materials from which they have been derived, namely, the materials of the typical deep-weathering profile, currently exposed by post-deep-weathering dissection. Stephens (1946), Gunn (1967), and others have made similar observations from other parts of Australia (Fig. 4.3).

But many mature soils have also been shown to reflect the parent material. For example in Scotland, brown earths have been observed to form on basic igneous rocks and podzols on acid igneous rocks, under identical weathering and pedogenic conditions. The presence of calcium has also been observed to cause the flocculation of aluminium, iron, and

SOIL FORMATION II

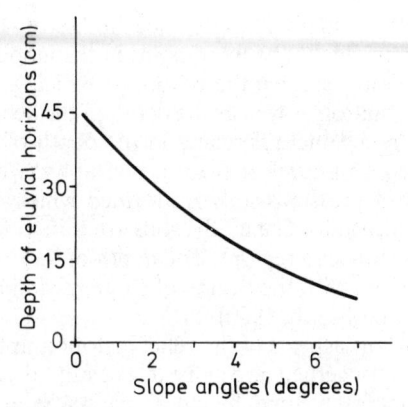

Fig. 4.3   Relationship between slope steepness and depth of eluvial horizons

humus, thus inhibiting movement of soil solutions. Extremely porous sands tend to develop mature profiles more quickly than, say, clayey materials, in which gleying is very common. Finally, the soils of Nigeria serve as good illustrations of the significance of parent material. Discussed more fully in a later chapter, these soils tend to vary mainly between those formed on recent alluvial and coastal deposits, older sedimentary rocks, basement complex rocks, and volcanic rocks (see Chapter 8). Similar examples of the significance of the parent material exist for many other countries. Therefore, although parent material does not generate sufficient energy to effect soil formation, it does exert very strong and in some cases decisive influence on the process of soil formation. Not only does it provide the soil mineral matter which constitutes almost 50 per cent of the soil body by volume and a lot more by weight, but its nature, characteristics, and composition are very important determinants of the soil. The parent material determines to a large extent the most important properties of the soil such as colour, texture, structure, and reaction (see Chapter 2).

TOPOGRAPHY AND SOIL FORMATION

Unlike climate, organisms, and parent material which either contribute materially to or participate in the formation of the soil body, topography is important essentially as a factor influencing these other factors. For instance, relief is a major control on the redistribution of both the mass (parent material, organic matter, soil solution, etc.) and the energy involved in soil formation.

The very nature of the recognized relationship between topography and soils has undergone many changes since this factor was first identified and studied by Dokuchaev. The main point in the association is the fact that soils develop on a surface the nature of which greatly conditions the groundwater situation as well as the depth of the accumulated debris. Thus Dokuchaev noted the differences in the depth of soil profiles due to undulation, even on a level or plain topography; while other people have found that in depressions soils are formed which differ from those formed on higher locations. Generally, soils are thin in hilly regions, and are much deeper in lowland regions. The depth of the soil A horizon has in fact been correlated with the degree of slope, giving a parabolic relationship (Norton and Smith, 1930).

The most important aspects of the relief or topographic factors, as far as soil formation is concerned, are slope (especially its angle and length), site or location along the slope, height above sea level or altitude, and aspect. Each of these acts both directly and indirectly to modify the processes of soil formation. For example, slope, as noted above, is an important factor of soil depth, especially the depth of the topsoil or A horizon. Generally, soil erosion is rapid on steep slopes, leading to thin topsoils, and slow on gentle slopes, leading to deep soils. Fig. 4.4 is an illustration of the direct effect of the slope factor from a part of Australia. It shows that soil depth decreases as slope angle increases, in what looks like a geometric progression.

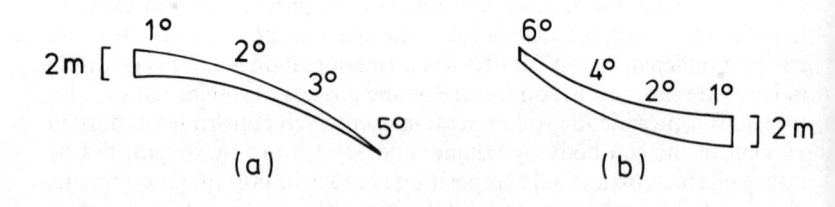

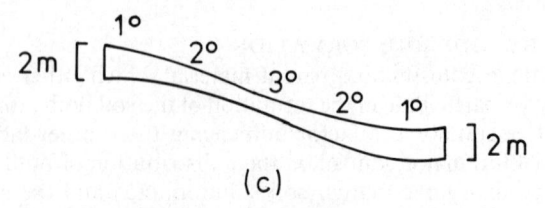

Fig. 4.4 Changes in solum thickness with slope angle (Mudgee area, Australia)

SOIL FORMATION II

With regard to position or location along valleyside slopes, it is important to note that soils are washed down from the upper to the lower portions of a slope segment, so that, theoretically at least, the depth of the soil is a function of distance from the interfluve to the valley bottom. This condition has been observed in many places, where the soil depth increased downslope (Fig. 4.4a). However, this condition is reversed in Fig. 4.4b; while Fig. 4.4c shows a complex relationship. In all these cases, it is the effect of slope angle that appears to outweigh that of slope position.

Slope position is, however, important in other respects. Leaching is pronounced on the upper slope, as fresh water continually penetrates the soil. The percolating water carries with it the salts dissolved from the upper-slope soils. The cations and anions in the lower-slope soils become less soluble and more concentrated, leading to high pH at these (lower-slope) positions. Lower-slope soils are also generally wetter than upper-slope soils, which condition favours the hydration of the oxides at these (lower-slope) positions, and the production of yellow rather than red or brown soils. By contrast, the upper-slope soils are better drained and so are reddish in colour. Mottling in soils is facilitated by fluctuating groundwater level and repeated seasonal or periodic waterlogging conditions, mainly at the lower-slope positions. Finally, slope aspect can significantly affect the amount of solar warming, so that different soils develop on sun-facing slopes as compared with slopes backing the sun. Aspect is especially important in high-latitude regions (cf. Fig. 4.5).

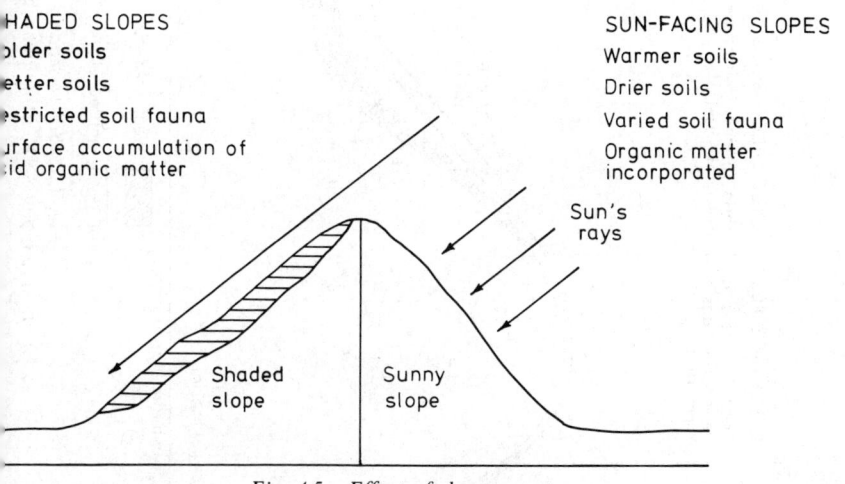

SHADED SLOPES
Older soils
Wetter soils
Restricted soil fauna
Surface accumulation of acid organic matter

SUN-FACING SLOPES
Warmer soils
Drier soils
Varied soil fauna
Organic matter incorporated

Fig. 4.5 Effect of slope aspect

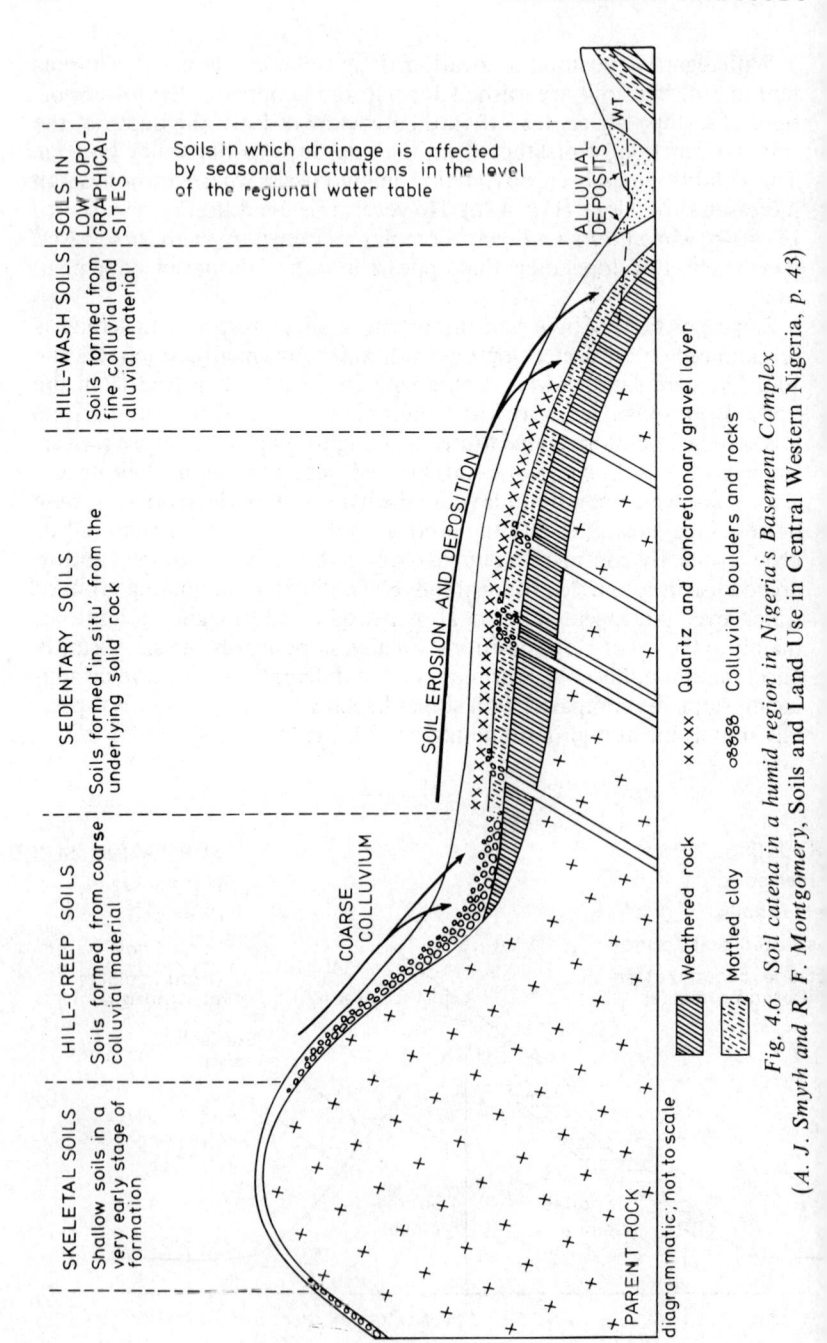

Fig. 4.6 Soil catena in a humid region in Nigeria's Basement Complex
(A. J. Smyth and R. F. Montgomery, Soils and Land Use in Central Western Nigeria, p. 43)

The variations being described have been designated the *catena* effect. This term, which was introduced by Milne (1935), refers to the sequence of soils found along many valleyside slopes (Fig. 4.6). The typical catena occurs in drainage basins (on valleyside slopes) of homogeneous geology, but may be complicated by lithological variations. Many examples are therefore in existence. For example, the variation in the soils of dissected duricrust regions characteristically illustrate the catena idea but with a difference. As discussed earlier, the variations observed from hilltop to valley bottom tend to reflect parent material.

The effect of altitude on soils is mainly through the action of climate. Altitude lowers temperature and increases precipitation. The result is seen in the zonation of soil regions up the sides of lofty mountains (Fig. 4.7, p. 62). Depending upon location and climate, the zonation pattern will vary, e.g. as between humid tropical and humid temperate regions.

The topographic or geomorphic factor in soil formation is closely related to the mode or process of slope evolution. Depending upon which of the geomorphic processes of peneplanation, pediplanation, etchplanation, etc. operated in an area, the soil pattern from valley bottoms to hilltops will differ significantly, because of the temporal and spatial complications involved. It is here that the study of soils should greatly assist geomorphology and vice versa. The nature of the soil on slopes should be able to throw some light on whether the slopes retreated and are perhaps still retreating, or whether the slope process is that of decline, while the rate of slope evolution can be used to determine the age of the soils and vice versa.

The Time Factor in Soil Formation
Like the other factors considered earlier, the role of this factor has been discussed in various ways. Originally it referred to the age of the country rock, but this has little or nothing to do with the soils of an area. The time factor has also been considered in relation to the duration of the pedogenic process from the time of its inception to the time the mature soil is formed. Definitely, the time necessary for soil formation varies widely. Soil horizons may become apparent within a short season, although such horizons usually represent the results of only a few of the processes of soil formation, e.g. the leaching of clays and other hydrous elements from a heap of weathered debris. More often than not, mature soil profiles form after about twenty or thirty years at the earliest, while some soil profiles need about a thousand or more years to form.

The concept of the mature soil has been widely discussed. In this volume, it will refer to the state in the evolution of the soil, where little or no noticeable change occurs in the profile. This means that the depth, of both the entire soil profile and its component horizons, the colour, pH, texture, etc., remain constant, despite the fact that soil is continually

Fig. 4.7 Mountains and soil zonation

metres	
4700	Permanent snow
	striated soils
4300	
	shallow mountain meadow soils
3500	
	peaty or brown alpine soils
	mountain meadow soils
3000	gleys
	grey ando soils
	young yellow / brown tropical soils
	humic ando soils
2400	hydromorphics (fog belt)
	humic mountain soils
2000	podzols on gravel
	pale ferrisols
	humic ferrisols
1500	brunizem on clay
	margalitic & gilgai
	red / yellow latosols
1000	
	ferrallitics
750	level of Lake Tanganyika
	KIVU
Sea level	

Fig. 4.7   Mountains and soil zonation

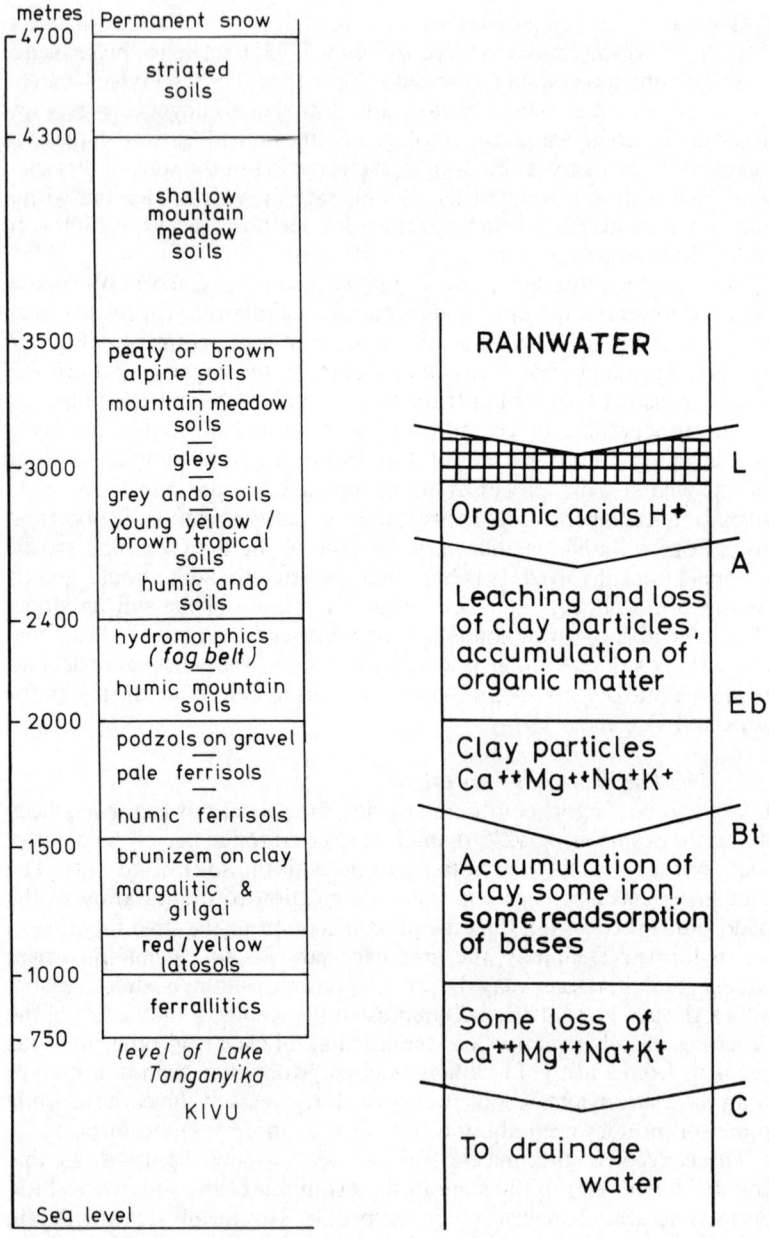

Fig. 4.10 The process of leaching

RAINWATER	
	L
Organic acids H$^+$	
	A
Leaching and loss of clay particles, accumulation of organic matter	
	Eb
Clay particles Ca^{++}Mg^{++}Na$^+$K$^+$	
	Bt
Accumulation of clay, some iron, some readsorption of bases	
Some loss of Ca^{++}Mg^{++}Na$^+$K$^+$	
	C
To drainage water	

Fig. 4.10   The process of leaching (See p. 66.)

being formed and destroyed. This concept of the steady state, or state of dynamic equilibrium was first described by Nikiforoff (1942, 1943), and has been further elaborated since. One aspect of the concept is where soil colour is shown to be constant. Certain soil-forming processes are faster than others. The leaching of clay and other bases and changes in colour and pH appear to occur within relatively short periods, e.g. within about five hundred years. However, observations have shown that the formation of clays and soil colloids in certain environments takes much longer, often running to thousands of years.

The time factor in soil formation may therefore be viewed in terms of pedogenic progression and the formation of mature soil profiles. We cannot set a time limit, except when other factors, e.g. parent material, climate, vegetation, etc., are held constant. Therefore counting years is of little use. Instead, we may consider time in relation to the stages in the development of the mature soil profile, as shown in Fig. 4.8 (p. 64).

The time factor in soil formation also presupposes the concept of periodicity and polygenesis. The environmental factors of soil formation do change, and with them the original soil may change character. A gradual change, e.g. a progressive increase in rainfall or a slow drop in temperature, may cause an existing soil to be transformed into a different soil type. This is the idea of polygenesis which refers to the derivation of one soil profile from another under a gradual change in environmental conditions. It is different from monogenesis, which refers to the formation of a soil directly from fresh parent material, under uniform conditions.

Changes in environmental conditions may, however, be catastrophic or marked. This may lead to the truncation (stripping) or burial of an original soil profile, or to both. Indeed studies made in many areas show that periods of soil formation and soil destruction have occurred one after the other, leading to complex and compound soil profiles, described as K-cycles. The topic of soil periodicity is a very popular one, which has been discussed widely in the literature. Among the most quoted examples are buried soils, in which case two or more distinct soil profiles occur one beneath the other, and new soils are formed on the truncated materials of an original soil profile or paleosol (Fig. 4.9a). This occurs in the Sydney district of New South Wales, where most soils are now formed on the materials of a fossil soil (deep-weathering) profile.

The major causes of change in the soil-forming environment are climatic and geologic. A change from a humid to a dry climate is likely to be accompanied by the stripping of soil profiles in some (elevated) areas and their burial in other (low-lying) areas. The arrival of humid conditions means further soil formation both on the truncated profiles and on the deposited material. Uplifts, diastrophic movements, etc. also disrupt the process of soil formation.

ESSENTIALS OF SOIL STUDY

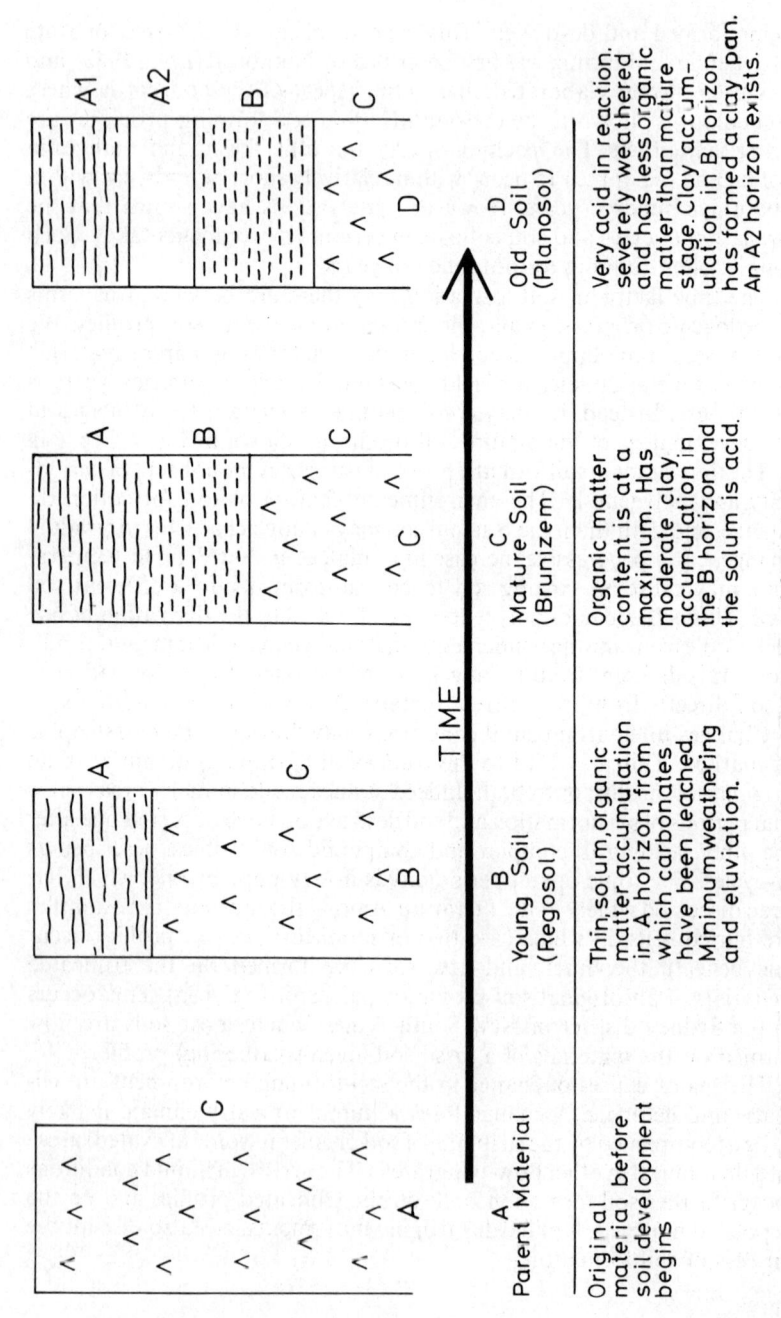

A Parent Material

Original material before soil development begins

B Young Soil (Regosol)

Thin, solum, organic matter accumulation in A horizon, from which carbonates have been leached. Minimum weathering and eluviation.

C Mature Soil (Brunizem)

Organic matter content is at a maximum. Has moderate clay accumulation in the B horizon and the solum is acid.

D Old Soil (Planosol)

Very acid in reaction, severely weathered and has less organic matter than mature stage. Clay accumulation in B horizon has formed a clay pan. An A2 horizon exists.

Fig. 4.8 *Time factor in soil formation*

SOIL FORMATION II

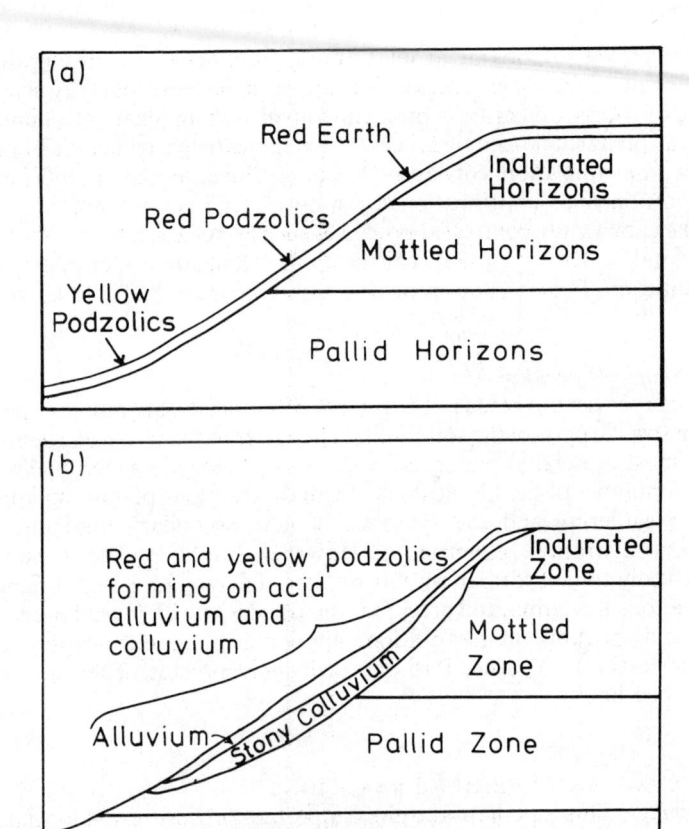

Fig. 4.9 Topographic dissection and soils

The Processes of Soil Formation

These are the processes which act both singly and collectively to modify saprolite, or essentially inorganic (dead) rock debris, to produce the living soil. The processes in question may be discussed in various ways. They may be considered in terms of whether they involve: (a) the accumulation of solid and also colloidal material, e.g. the process of rock weathering (see Chapter 3); or (b) the differentiation of horizons by addition, removal, and transfer of materials and energizing solutions. Additions to the profile include mainly organic matter from the biosphere; removals concern the soluble substances, e.g. the salts and carbonates; while humus and sesquioxides are generally transferred from one part of the profile to another. Transformations also occur, e.g. of primary organic matter to humic acids, and of primary minerals to secondary ones.

Another way of looking at soil-forming processes is to consider them at different levels. Some processes are simple, in the sense that they involve a single process; others are complex and involve an amalgam of a number if not all of the single, simple processes. Examples of the former are humification, mineralization, eluviation, leaching, illuviation, ammonification, nitrification, and denitrification, to list only a few; examples of the latter are the climatically controlled processes such as podzolization, lateritization (and/or ferralitization), calcification, salinization, gleization, and solodization. These various processes are summarized in Table 4.2, p. 70.

The Simple Processes
These processes are noted not only for their simplicity, but also for operating in specific parts of the soil profile. They refer to the physical, chemical, and (most especially) biological reactions or processes which make the soil. A number of the physical and chemical reactions pertain mainly to rock weathering and the synthesis of new, secondary minerals, e.g. solution, hydration, reduction, etc. However, a number of others pertain specifically to soil formation and horizon differentiation, e.g. leaching, eluviation, illuviation, and precipitation. Besides, the biological processes (e.g. humification, mineralization, ammonification, nitrification, and denitrification) are crucial to the pedogenic process. They are now described briefly.

PHYSICAL AND CHEMICAL REACTIONS
Leaching This is perhaps the most important process of soil formation, especially with respect to horizon differentiation. It operates to some extent in all soils and is fundamental to all the more complex soil-forming processes such as lateritization, podzolization, and calcification. By definition, leaching means the removal in solution of constituents from the soil. As such it operates mainly in humid environments and is least effective in dry areas. The materials involved in the leaching process are mainly the salts and the carbonates, the readily soluble minerals. Clay is also translocated during the processes of leaching (Fig. 4.10, p. 62).

Leaching is both beneficial and destructive to soil. It is fortunate that the most soluble and the most readily leached substances are also the most dangerous to plants—e.g. sodium chloride—to the extent that they are very toxic. However, excessive leaching robs many soils of essential plant nutrients, which can only be extracted by plants in the soluble form, the very form in which they are leached from the soil. Leaching may involve the complete loss of nutrients and other soil materials from the profile, e.g. loss of silica during lateritization or of base elements to drainage water in humid environments. Alternatively, it may involve the removal of constituents from one part of the soil to another. For example, in some soils, clays, sesquioxides, and decomposed organic matter are mobilized and translocated from the A to the B horizon, as in the process of podzolization (see below).

The constituents which are leached from soils may be divided into ions and colloids according to their size and solubility. Ions, which are either positively (cations) or negatively (anions) charged bodies, exist only in the soil solution; their concentration therefore depends on the solubility of the source material. Ions are smaller in particle size than colloids and so are more likely to be completely removed from the soil profile.

Eluviation and illuviation Closely related to the process of leaching but not identical to it are the physical processes of eluviation and illuviation. Both processes refer to the movement of soil material, in solution or in suspension, from one place to another within the soil. They operate mainly in areas where there is an excess of precipitation over evaporation, i.e. the same places where leaching takes place. However, while leaching may involve total loss of substances from the soil, eluviation refers only to translocation within the soil. Eluviation may also be vertical (upward or downward) or lateral (sideways), depending upon the direction of movement of soil water. Eluviation also refers especially, but not exclusively, to the movement of colloids.

Horizons that lose material through eluviation are referred to as eluvial while those that receive the material are called illuvial horizons. Illuviation therefore refers to the deposition in a soil layer of colloids, soluble salts, and small mineral particles which have been transferred or leached out of an overlying or underlying soil layer. In relation to the soil profile, eluviation takes place in the A horizon and illuviation in the B horizon.

Eluviation and illuviation may operate in an upward direction in the soil profile, e.g. in the capillary movement and precipitation of salts, etc., at or near the soil surface. This situation obtains in a dry environment, where evaporation is marked. It is the basic process of calcification, just as leaching is basic to podzolization. Where it occurs, the top part of the soil (A horizon) will be the illuvial horizon and the lower part (B horizon) the eluvial horizon.

BIOLOGICAL REACTIONS
Humification This is the process by which organic matter is decomposed and new organic complexes are synthesized to form humus or the organic matter component of the soil. It is perhaps the most important process, since without it there cannot be any soil. It is also partly responsible for the appellation 'the living soil', mentioned earlier in this volume (Chapter 1).

The rate of accumulation of the organic residue—e.g. dead leaves, twigs, branches, tree trunks, herbaceous vegetation, remnants of dead animals, etc.—and the general characteristics of the resulting material (i.e. humus) are all principally controlled by climate. This means that as climate changes imperceptibly from one type into another, the intensity of humification changes and a different set of conditions takes over. This is represented in Fig. 4.11. According to this diagram, at least four pedogenic regions may be identified. First, in the humid tropics and subtropics, with constantly high temperatures and heavy rainfall, conditions are ideal for microbial activities and thus for the break-down of the natural organic matter and of the synthesized humus complexes. In this region, therefore, organic matter is rapidly 'burnt up' and so does not accumulate to any extent. The burning out of the humus is often referred to as mineralization.

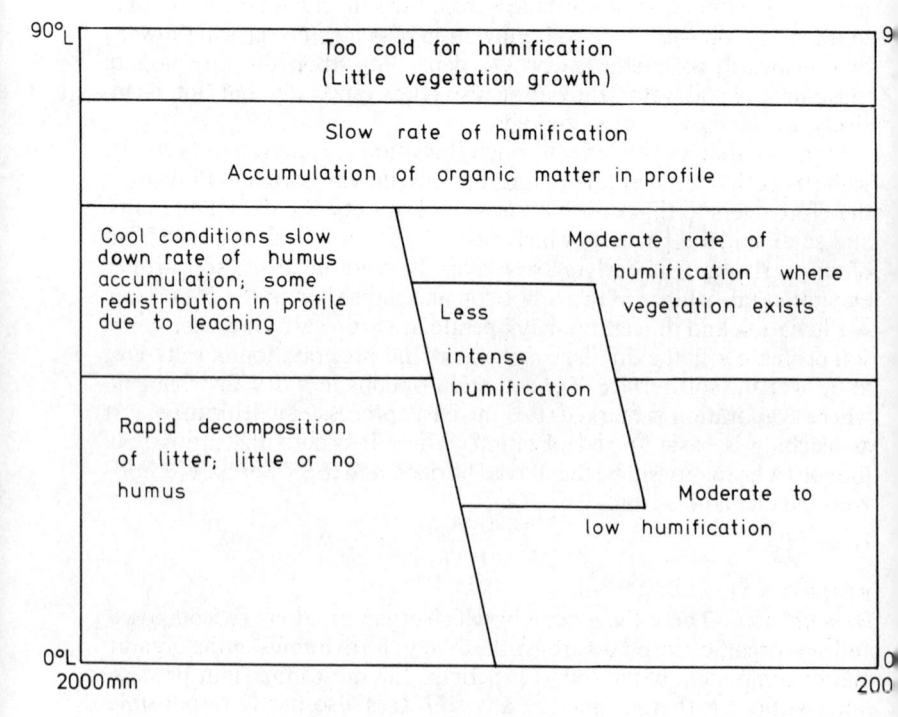

Fig. 4.11   Climate and the process of humification

Second, in the cool temperate regions, the process of humification slows down. This is partly due to the shortening of the favourable season for microbial activities, and partly due to the effects of lower temperatures which are not conducive to oxidation and the humification process. Consequently, the organic matter residues tend to accumulate mostly unchanged for much longer than in the humid tropics. The result is the production of humic acids which are important in the process of leaching. This area is therefore noted for the intensity of the leaching activity, as evidenced by the leached A horizon of the podzols.

A third humification region is the cold and wet areas of the cold temperate and arctic regions. Here the process of humification is virtually nil. Anaerobic conditions are widespread, temperatures are constantly low, and microbial activity is negligible. The result is a deep Ao layer, and the widespread occurrence of peat. Humification is similarly negligible in hot deserts on account of scanty plant life and the usually excessively dry conditions.

Finally, in the grassland regions, the humus or decay accumulation layer blends in with the horizon of eluviation. The true Ao horizon is represented by a mat of dead vegetation or the sod of the meadow in the humid regions, and of the prairie in the less humid, partly arid regions. These four regions are identical to those of the major soil-forming processes described below.

Apart from climate, the type of vegetation also affects the humification process. Generally, woody plants contain large quantities of lignins, waxes, and resins, which are highly resistant to decomposition. They are also rich in tannins and other acid-producing substances which tend to subdue the activities of bacteria and stimulate fungus development. By contrast, in the grasslands, the process of humification is intense, albeit at the expense of the mineralization process. The result is the accumulation of organic matter.

Mineralization This is the process by which the organic mineral elements in the soil organic matter are released and converted into inorganic minerals. The process involves the decomposition of organic matter to its fundamental constituents, i.e. O_2 and other gases, water, and minerals. The process takes place simultaneously with humification, mainly in the Ao layer, but also at any other part of the soil profile where organic matter is present. Unlike humification, however, which tends to increase the acid content of the humus complex, mineralization tends to decrease the acid and to increase the base content of the soil, thus tending to make the soil alkaline.

Other biological processes The other processes—ammonification, nitrification, denitrification, etc.—are all connected with the process of nitrogen fixation and nitrogen cycling in the biosphere. The nitrogen content of most soils ranges from 0·2 to 0·4 per cent, the major part (92–96 per cent) being insoluble organic combinations inaccessible to plants.

The most important aspect of the biological process, especially in relation to soil fertility, is perhaps the fixation of nitrogen by legumes, otherwise called *symbiotic fixation*. Some legumes are able to extract nitrogen from the soil atmosphere and incorporate it into their system. This nitrogen is made available to other plants when the host plant dies and is ploughed back into the soil.

The Complex Processes

The processes of soil formation are commonly seen as amalgams of many of the above-described simple processes. This is quite understandable because none of the simple processes operates on its own; the normal thing is for a number of them to operate together, depending upon the climatic and biotic factors.

The most commonly discussed processes are podzolization, lateritization, calcification, salinization, gleization, and solodization, distinguished mainly on climatic grounds (Fig. 4.12). These processes come under the 'specific' as opposed to the 'fundamental' processes in Zakharov's system of classification (Table 4.2). They are not by any means mutually exclusive processes, although each tends to predominate in certain climatic environments. In general, gleying and podzolization may occur simultaneously in the same soil, while the laterite profile is what it is because of the process of gleying which operates under the influence of fluctuating groundwater. Moreover, as a result of changes in

TABLE 4.2

*The Processes and Reactions of Soil Formation**

	Processes			Reactions†	
	Fundamental	*Specific*	*Physical*	*Chemical*	*Biological*
1	Litter accumulation decay and formation of Ao (humus) layer	Humus formation, peat formation, vegetation growth	Infiltration through flow mechanical translocation	Solution precipitation, hydration	Humification, mineralization
2	Eluviation, formation of the A horizon	Leaching of bases, podzolization, lateritization, humus formation	Texture, structure	Oxidation, reduction, carbonation, silication, desilication, sorption, iron exchange	Ammonification, nitrification, denitrification, nitrogen fixation, etc.
3	Illuviation, formation of the B horizon	Accumulation of lime, clay, iron and aluminium			
4	Differentiation in the mass of the various horizons	Soil solution, soil maturation, aeration			

* Modified from Zakharov, 1946, p. 256
† The reactions take place together and in varying combination at all stages.

environmental situations, an existing soil may be completely changed. Thus we may have a podzolized laterite or a lateritized podzol. The soils of a large part of south-east Australia illustrate the situation in which the present-day podzolic processes are operating on fossil laterite and/or ferricrete profiles and their truncated forms.

The processes in question are also in the main rather hypothetical. Most of them operate very slowly, taking thousands of years to complete. They are therefore not readily observable. Some aspects of these processes—e.g. of podzolization—have been reproduced under laboratory conditions, but the various activities involved in this and the other processes remain mostly inferential, deduced partly from the hypothetical model and partly from their end-products, e.g. the soil profiles themselves.

It is important to re-emphasize that, as in all natural phenomena, boundaries are mostly arbitrary. The various processes have core regions where they operate optimally, e.g. podzolization in the coniferous forest regions or the taiga. From such core regions the processes change their character gradually and imperceptibly as the factors—particularly climate and biota—change. These notwithstanding, we shall treat the processes separately, if only to facilitate easy comprehension.

THE PROCESS OF LATERITIZATION

This is the process which forms laterites and lateritic soils. These soils, which are also variously called latosols, oxisol, and ferruginuous tropical soils, are described in a later chapter (Chapter 8). We are here concerned only with the process itself. It needs to be stressed at this point, however, that the proliferation of terminologies, some of which are hardly called for, is to be detested. An example of such terms is latosol, which does not seem to add anything to, or have any advantage over, the original term, laterite. We are thus adopting the original terms (i.e. laterite, lateritization, also laterization), although we do not object to some of the other terms such as ferrallization being used in some instances.

The process of lateritization operates mainly in the warm to hot humid regions, particularly in the areas classified by Koppen as having A climates. These are the equatorial (Af) climate, the tropical (Aw) climate and the monsoon (Am) climate. Besides the climatic and also the implied vegetation factors, the topography is another important controlling factor of this process. Topography is particularly important in relation to the groundwater situation, as well as to the depth of the effective soil.

The following characterize the lateritization process:

(a) Given the humid tropical conditions and perhaps also a long uninterrupted period of rock weathering and soil formation, deep and intensely weathered parent materials occur.

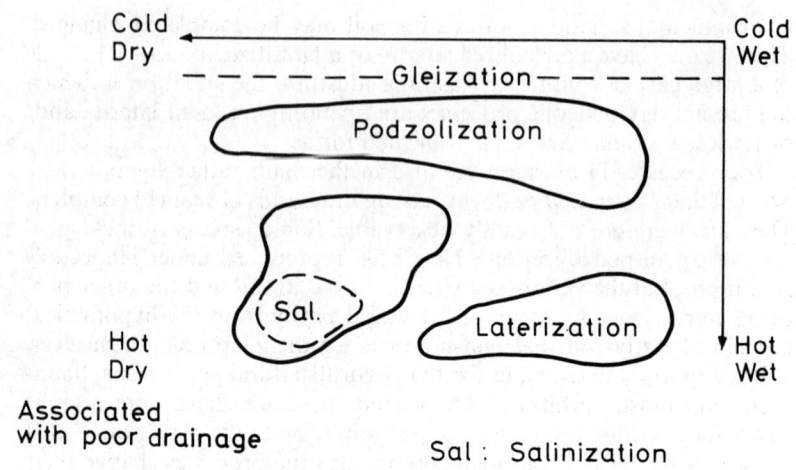

Fig. 4.12 Pedogenic zones

(b) The leaf fall from the luxuriant forest and its rapid decay keeps the bases in rapid circulation between the soil and the vegetation.

(c) Mineralization of organic matter is complete and rapid, while the intense chemical weathering of rocks releases bases from the mineral fragments still remaining in the soil. The result is a mildly acidic to mildly alkaline (pH 5–8) soil solution. This is conducive to the process of desilication, or the removal of combined silica in preference to the iron and aluminium which gradually accumulate and oxidize to the sesquioxides, under conditions of free drainage (Fig. 4.13b). The process of desilication lies at the heart of the lateritization process.

(d) Lateritization also involves the alteration of the high-silicate clays (montmorillonite, illite) into low-silicate clays, e.g. kaolinite. Alteration may, of course, progress to the point where only a residue of kaolin and iron and aluminium sesquioxides, in various degrees of hydration and having small inclusions or impurities, are left.

(e) Finally, depending upon the groundwater situation, the process of lateritization may result in the formation of a stratified or an unstratified profile. In the former case a marked but fluctuating groundwater table facilitates the formation of an indurated zone above the water table, a mottled zone in the zone of fluctuating water table, and a pallid zone in the zone of complete and permanent saturation. This is the typical laterite profile, also called the deep-weathering (duricrust) profile (Faniran, 1970). The unstratified profile has been called various names including the 'kaolinized' profile (Wright, 1963) and krasnozem (Corbett, 1969). The former connotation is obviously wrong, since a kaolin zone is also associated with the stratified profile.

THE PROCESS OF PODZOLIZATION

Prevalent in the cool humid parts (e.g. Koppen's C and D climate), this process produces the podzols and podzolic soils. Just as the process of lateritization is most intense in the formation of the tropical laterite profile, the process of podzolization is most severe in the formation of the podzols. Here the profiles are more distinct both in appearance and in their physical and chemical properties than in the podzolic soils. Indeed, the ideal soil profile, which is also taken as the typical soil profile, is the podzol, because of the clear-cut development of the horizons and parts thereof.

The word podzolization is derived from the Russian term podzol, which means 'ash underneath'. Like lateritization, the principal operating process is leaching, but with a difference. Because of the high acid content of the soil solution (pH 3–4), the leaching, especially of the bases and soil colloids, is much more intense than in any other soil. The outcome is as follows (Fig. 4.13a):

(a) Soluble bases (metallic cations) are displaced from the soil colloidal interfaces and removed from the solum by groundwater.
(b) Aluminium and iron sesquioxides become mobile and are transported downward in the soil, thus making the A_2 horizon somewhat lighter in colour than it was previously. Unlike in the case of lateritization, however, silica is not so affected and so tends to remain in place.
(c) Inorganic clays tend to be peptized and susceptible to dispersion and downward movement.

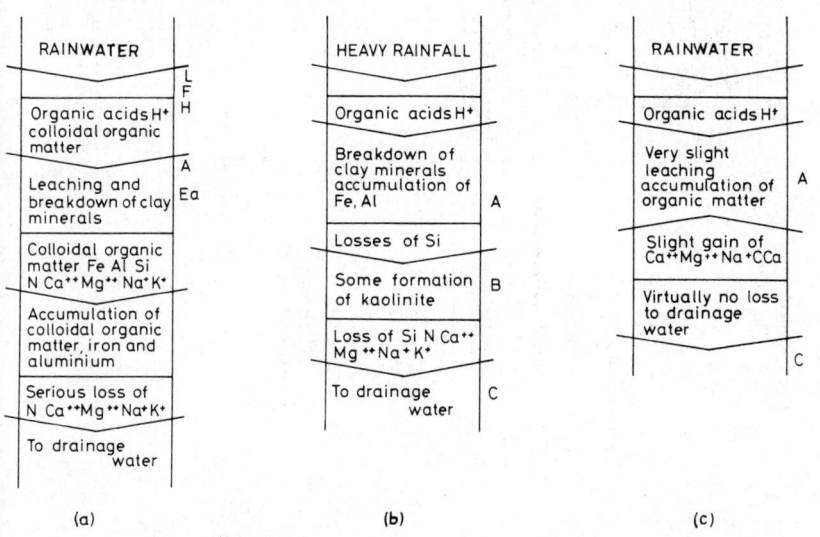

Fig. 4.13 Processes of soil formation

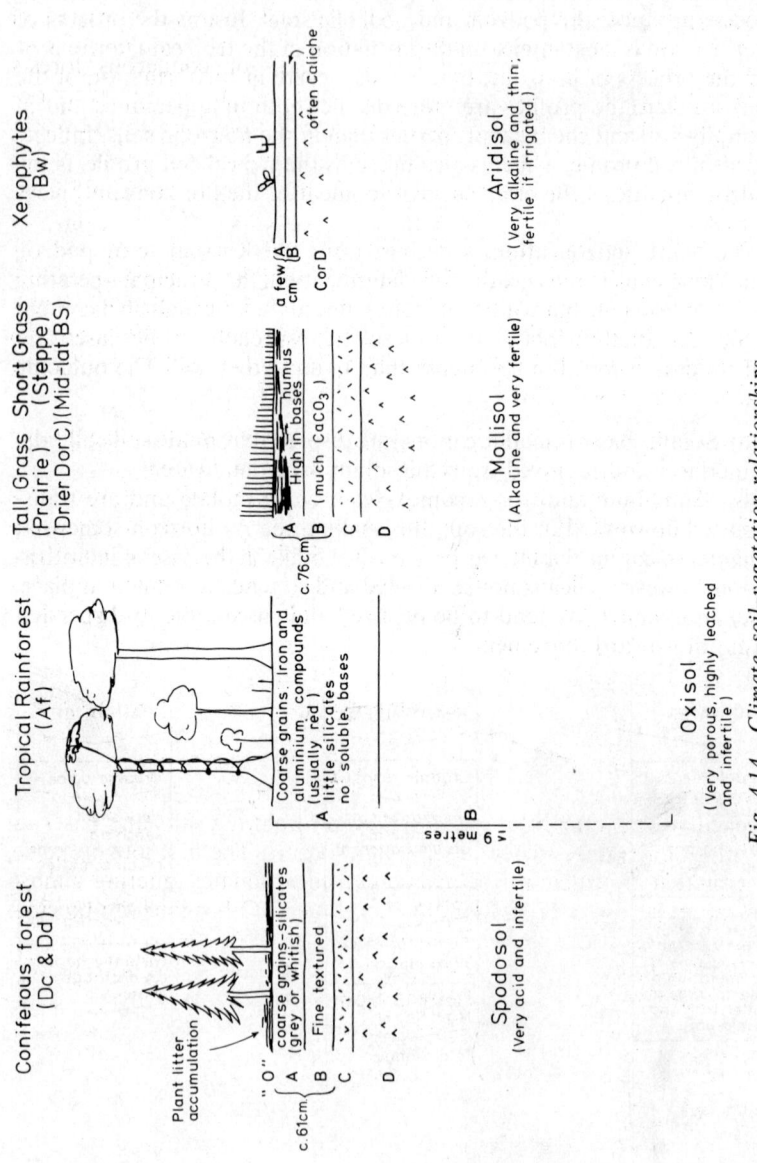

Fig. 4.14 Climate–soil–vegetation relationships

(d) An illuvial B horizon tends to develop at depth, as a result of the deposition of the sesquioxides, clays, bases, and humus, which primarily results from reduction in soil acidity.

Podzols are best developed beneath heath or coniferous forests (Koppen's Dfc–Dfb climates), while podzolic soils are more widely developed in the temperate latitudes under a wide range of vegetation and climatic types.

THE PROCESS OF CALCIFICATION

While the two processes considered already operate in humid regions, the process of calcification is characteristic of low-rainfall areas, especially in the continental interiors. As such, it is characteristic of the grassland regions, including the savannas, but especially the temperate grassland areas.

The process involves the formation of a calcic[1] horizon (usually within the B or C horizon) and a mollic horizon or a soft, friable surface horizon containing a high percentage of organic colloids, well saturated with calcium ions. Indeed the prevalence of lime in these soils caused them to be collectively called pedocals, or soils of calcium. These soils are today called mollisols, among other names, and the chernozem is perhaps the best example.

Leaching is not a prominent feature of this process, and although some downward movement of solutions and materials takes place, these are not normally removed from the profile (Fig. 4.13c). This is mainly because evapotranspiration exceeds precipitation in the dry climates where this process operates, which means that there is some net loss of soil water as it passes through the soil. As evaporation takes place, soil solutions have an increasing concentration of salts and bases. A point is then reached at which the solution can no longer hold the salts and bases as dissolved ions; this is when precipitation occurs. The first ions to be so precipitated are calcium and magnesium bicarbonates, altering almost immediately after precipitation into carbonates. Other ions are precipitated later, e.g. calcium sulphates. Depending upon the abundance of calcium in the original rock, and also on the duration of the calcification process, a crust may be formed, similar to the indurated zone of the laterite profile and usually called calcrete (cf. French *croûte calcaire*).

THE PROCESS OF SALINIZATION

With increasing dryness, the process of calcification is modified into that of salinization, which essentially involves the enrichment of a soil with

---

[1] See Chapter 7 on Soil Classification.

salt. As in the case of calcification, enrichment is accomplished by the evaporating soil solutions. Salts in solution are drawn upwards by capillary action and are then deposited at the soil surface, as the water is lost to the atmosphere. The soils so formed therefore typically have surface incrustations of salt. They are called white alkali soils or solonchaks and have marked salic horizons.

Related to this process are the processes of solonization and solodization. Indeed, there is a succession of soils from solonchaks (saline soils) through solonetz, solodized-solonetz, to solodic soils. This succession illustrates the delicate balance between the process of salt accumulation and that of salt removal in the soil; in other words between the process of capillary precipitation and of leaching. When salt accumulation predominates over all other processes the resulting soil is a solonchak soil. But when salt accession is fairly strong, e.g. when there is some illuviation of free salts to the lower parts of a soil profile, the resulting soil is a solonetz. When the salts are redistributed through the entire soil profile, i.e. when leaching is more marked, a solodized-solonetz soil is formed. Finally, the salt may be removed from the profile by leaching and the profile strongly differentiated. The result in this last case is a solodized soil. Thus, both solonetz and solodized-solonetz soils may be called alkaline soils, while solodic soils are very much like podzolic soils and so are more acidic.

THE PROCESS OF GLEIZATION
The processes described so far operate mainly in well-drained areas. However, there are many areas where drainage is not so good and soils have to form somehow under impeded drainage conditions. The soils formed under such conditions are grouped together by pedologists as hydromorphic soils.

Gleying can occur in any climatic environment, i.e. within zonal soils. This may be due to topographic effect, e.g. swampy low-relief areas, valley bottoms, etc., or to geological influence, e.g. poorly drained, clayey soils. Gleying may also be due to climate, for instance in the permafrost regions of the cold regions. In this last region the permafrost provides the impervious horizon, usually at very shallow depths, while the top soils are usually saturated by ice-melt water. The soil profile therefore has little chance of developing, and so is in a perpetual state of youthfulness. The name given to this type of soil in the American 7th Approximation is 'inceptisol' (see Chapter 7).

Gleying outside the arctic polar regions is slightly different, essentially as a result of the effect of warmer temperatures. For one thing, mottling is characteristic of soils in such areas especially where periodic drying takes place. Nevertheless, it is possible to generalize the discussion here,

SOIL FORMATION II

and describe the process in terms of surface-water and groundwater gleying. An impervious horizon within the soil profile causes surface-water gleying, while water rising in the soil from an impervious horizon beneath the soil produces groundwater gleying. With complete saturation, the colour of the soil so formed is frequently homogeneously grey or greyish-white, while occasional drying up of the soil, as observed earlier, produces mottles of rust-coloured ferric oxide. Mottling is a prominent feature of most hydromorphic soils.

Finally, extremely poor soil drainage conditions may result in peat formation in addition to gleying. The peat so formed may be acid or neutral to mildly alkaline. The acid moor peat develops on upland areas where high rainfall results in the leaching of all the bases. By contrast, the fen peats develop in waters liberally supplied with bases. Unlike the acid moor peat soil, fen peats are potentially rich if effectively drained.

Chapter 5

SOIL SURVEY TECHNIQUES AND PROCEDURES

Soil is a very important natural resource, regarded by many as the chief element of land. Soil has played a major role in the history of mankind, influencing the agricultural productive capacity and so the rate of growth and the survival of populations, the distribution of settlements, and movements of trade and people all over the world. In order to meet the needs of an ever-increasing population it is essential that the soil be put to the best use. In order to do this it is necessary first to know the different types of soils present in an area, their properties, and relationships in space. Soil survey is concerned with the inventory, classification, and systematic mapping of soils. It provides data on the composition and properties of soils, their arrangement in space and the role of the different soil-forming factors, and on the use of the resource. Apart from satisfying purely scientific and academic interests, the practical purpose of soil surveys is 'to provide a systematic basis for the study of crop and soil relationships with a view to increasing productivity and to help in soil conservation and reclamation' (Stephens, 1953). According to Gerasimov and Glazovskaya (1965), a soil map, with an explanatory text, provides the most precise and complete description of natural soil properties within a territory, on the basis of which one may determine the lands most suitable for various utilizations in rural economy, ensure the best agricultural organization of the territory, and select the proper farming practices and meliorative measures with a view to enhancing soil fertility, or for purposes of conservation.

The assessment of the quality of soils and their capability for different types of land use is therefore part of the work of the soil suvey. In this respect soil survey interpretation assumes great importance in that it translates the results of detailed soil research and field investigations into terms used by the community.

Soil Survey Organizations

Almost every country in the world now has a soil survey organization concerned with the study, mapping, and conservation of soil resources. This is the outcome of the phenomenal upsurge of interest in the environment and world natural resources which has taken place since the end of the Second World War. In every country the establishment and development of a soil survey unit became a reality when soil research was directed more towards agricultural and other natural resources than towards purely scientific and academic pursuits.

The history of the development of soil surveys has recently been reviewed by Cruickshank (1972). Soil survey in one form or another began sporadically in different parts of the world, including Britain (for example, early nineteenth-century agricultural surveys), Russia (initiated by Dokuchaev about 1883), the United States (from 1863 onwards through the efforts of Hilgard, Shaler, and Whitney), and many European countries. But the establishment of organized soil survey units has occurred mainly during this century. The United States Soil Survey was established in 1898 under the direction of M. Whitney. The Soil Survey of England and Wales was set up in 1939 under G. W. Robinson at Bangor, Caernarvonshire, and was transferred to its present site at Rothamsted Experimental Station in 1943.

The period following the Second World War saw the establishment of soil survey units in many countries, particularly in the less advanced areas in Africa, Eastern Europe, South America, Australia, and New Zealand. An important feature of the spread of soil surveys to these areas was that soil research became linked with environmental studies. Field scientists now look at the environment as an integrated whole, a system in which the different elements of rocks, water, soils, plants, and animals interact with one another and function together. In these countries soil survey work is being done not only by soil survey organizations as such, but also by land research units, e.g. the Commonwealth Scientific and Industrial Research Organization (CSIRO) in Australia, and the Land Resources Division (LRD) of the Ministry of Overseas Development, mainly in Africa but also in other tropical regions of the world.

A soil survey organization is not made up solely of soil scientists or pedologists. Its various activities provide job opportunities for different professions. A soil survey unit usually comprises the following categories of people:

1   Soil scientists and professional soil surveyors. These are the men really involved in soil research and field surveys.
2   Field assistants—usually unskilled or daily-paid labour to help the professional surveyor in the field. They clear the sites and prepare the soil profile pits and help in transporting equipment from point to point in the field.
3   Laboratory assistants. These are of three types: (i) those employed to assist in the laboratory analysis of soil properties; (ii) semi-skilled personnel engaged in photoanalytical work; and (iii) those engaged in collecting and processing data on soils.

4   Technical staff and machinists including motor drivers.
5   Cartographers and photogrammetrists.
6   Office and administrative staff—white-collar workers engaged in the administrative, financial, printing, and publications aspect of the survey project.

In each country the soil survey of the different parts is undertaken by numerous and diverse surveyors. Left to himself, each surveyor will probably adopt his own techniques, mapping units, and terminology. It is in order to avoid such a chaotic situation and to make the results of surveys carried out by different people in the various parts of the country comparable one with another that each survey organization produces a 'manual' of soil survey. The manual contains guidelines for the field survey, laying down standard procedure to follow and the scales of measurement to use in the description of soil morphological properties. The most notable of these manuals is the United States Department of Agriculture Handbook 18, *Soil Survey Manual* (1951). The manual in use in Britain derives from G. R. Clarke's *The Study of the Soil in the Field*. Most countries in Africa and elsewhere use the United States system, modified and adapted, if the need be, to their own local conditions. The FAO *Guidelines for Soil Description*, derived largely from the USDA *Soil Survey Manual*, has also been widely used. Also, each organization has a handbook on standard laboratory techniques of soil analysis. In fact in many places all soil samples collected during any soil survey project are analysed in a central soil-testing laboratory to ensure uniform treatment of the samples.

Soil survey organizations differ from country to country in form, degree of efficiency, and level of achievement. The factors which govern these include the expertise and skill of the available personnel, the organizational ability of the directors, the socio-economic environment of the unit, the level of technology, and the financial resources available.

Stages in a Soil Survey Project
There are four steps or stages in a soil survey project (see Stephens, 1953; Areola 1971). These are:

1   *Preliminary Investigation*
The operations to be performed at this initial stage are as follows: (i) definition of the area to be surveyed and the practical tasks of the survey; (ii) review of the existing geographical data, in published and unpublished literature; (iii) planning and programming of field research; (iv) selection and procurement of instruments and equipment (see Gerasimov and Glazovskaya, 1965).

Every soil survey requires a certain amount of background research and field reconnaissance prior to the fieldwork. This is to gather information on environmental factors and to observe the broad soil patterns in relation to the geographic location of the area. Much of the information used to be extracted from topographical maps. Nowadays most soil surveys use aerial photographs in scanning the survey area and these sometimes replace much of the field reconnaissance that is necessary before the detailed field survey. Important features of the landscape such as geological formations, drainage lines, morphological units, and soil units which can be differentiated are all tentatively marked on the aerial photographs for subsequent verification in the field. The map and the aerial photograph are useful in delineating the traverse lines along which soils are to be examined in the field so that the survey covers, as much as possible, the variety of landscape and soil types in the area. The aerial photograph in particular is very useful in finding out beforehand the accessibility and ease of movement from place to place in the area.

2 Field Survey

This entails: (i) the examination and detailed description of soil profiles in the field; (ii) the selection of the soil-mapping units after a complete record of the location and morphological characteristics of every profile has been made; and (iii) the location of class boundaries. The three operations can be done concurrently in the field especially where soil–landscape relationships are easy to identify. In more difficult cases, however, soil classification for mapping purposes and the location of class boundaries have to await the detailed field study of soil profiles.

In order to achieve the main objectives of the fieldwork as stated above, three types of field observations are essential (cf. Buringh *et al.*, 1962): (i) classification observations, (ii) plotting observations, and (iii) special observations. Classification observations refer to the analysis of soil site and profile characteristics with a view to determining which soil-mapping unit occurs at a given spot. Plotting observations are made usually at the transition zones between different soil-mapping units to enable the accurate plotting of the class boundaries. Special observations are not necessarily made during the main fieldwork but may be made after the mapping and classification of the soils. Special observations are made to test special soil characteristics such as permeability, moisture-retentive capacity, stoniness, shear strength, etc.

Soil surveys can be classified into different types based on the scale of mapping, the number of field observations per unit area, the mapping procedure, the accuracy of the class boundaries, and the degree of uniformity within each mapping unit. The decisive factor in all these is the scale of mapping; every other thing is dependent on it. It is itself, to an extent, determined by the choice of the mapping unit. It is possible to recognize three main types of soil survey: (i) reconnaissance or general survey; (ii) semi-detailed survey; and (iii) detailed survey.

RECONNAISSANCE SURVEYS
Reconnaissance surveys are usually at scales of 1:100,000, 1:250,000, or less. They show broad soil groups at the higher categories of classification—e.g. association and catena—usually related to geological formations or topographical regions. In many cases the mapping and classification are done by the convergence of evidence method using geological and topographical maps, and, increasingly, aerial photographs. The amount of fieldwork is kept to a minium. The investigation of soil-profile characteristics may be limited to some specified horizon(s) or depth below surface and to some carefully chosen properties such as texture, structure, organic matter content, colour, and stoniness. In any case, most of the soil-profile inspection is done by augering along widely spaced traverses. These traverses more often than not follow the access routes, including roads and footpaths.

In a survey carried out in this way, class boundaries cannot be defined with precision, and very few specific statements can be made about the nature and behaviour of the soils in each mapping unit. However, a reconnaissance survey gives reliable information on the relationship between soils and other environmental factors such as slope and landforms, parent material, and vegetation.

The method is ideal for covering extensive areas of territory in countries which cannot afford the money or the time to engage in detailed soil surveys. Reconnaissance soil surveys of varying intensity have been carried out in most African countries, where the close relationship between the soils and the rocks and land forms in the continent has facilitated the application of the technique.

SEMI-DETAILED SURVEYS
Semi-detailed surveys are an advancement on reconnaissance surveys. The sampling density (that is, the number of sampling points per unit area) is greater; greater attention is paid to soil profile, as opposed to visual soil site characteristics; more soil properties are investigated and smaller, more uniform soil-mapping units adopted. Mapping is normally at the soil series level, at scales varying from 1 : 20,000 to 1 : 63,360.

A form of semi-detailed soil survey which is rapidly gaining ground with the development of aerial photography is pedological air photoanalysis (Buringh, 1954a and b, 1955; Jones, 1958) involving the investigation and evaluation of all land surface features which reflect differences in soil conditions. These include slope and landforms; drainage lines;

groundwater conditions; vegetation types; and land-use patterns. Probable soil units and soil boundaries are marked on the aerial photographs. At this stage no form of soil identification is attempted. Then, with the photograph as a base map checks are made in the field to ascertain the accuracy of the photoanalysis and to identify the types of soil within the areas classified on the aerial photographs.

Pedological air photoanalysis is commonly used in interpolating between two adjacent areas or in extrapolating from one area to another area with similar environmental conditions. The interpolation and extrapolation methods of soil mapping require the study of the relationship between the natural environment and the photographic imagery in one or more key areas. Soil conditions in other similar terrains can then be deduced from aerial photographs by analogy. The success of pedological air photoanalysis depends very much on the nature of the terrain and the extent to which soil conditions are reflected by visible landscape phenomena. The best results are obtained where the photoanalyst is not only skilled in photo-interpretation but also has a deep knowledge of soils, their formation and spatial relations, and the principles of their classification.

DETAILED SURVEY
Detailed surveys involve intensive soil sampling and analysis, within small geographic areas, to classify and map the soils at the lowest possible categories of classification such as the soil type and the soil phase. The method is favoured where it is important to know either the detailed pattern of the soils, or the degree of variation of a single soil property. The surveys are usually at scales of 1 : 10,000 and larger. They are laborious and time-consuming special purpose surveys in which there is very little reliance on deductions and inferences from maps or aerial photographs, most of the investigations being done in the field. Detailed surveys yield data on minor variations in soil properties and on soil/crop relationships. They are therefore employed when planning agricultural projects such as irrigation, plantation, and farm settlement schemes, and routine farm management practices such as liming and the application of fertilizers.

Soil surveyors approach the task of looking at the soils in the field and marking soil boundaries in different ways. Beckett (1968) has recognized three main approaches, namely, (1) grid mapping, (2) free survey, and (3) physiographic mapping. In grid mapping soil profiles are inspected and samples collected along regularly spaced traverse lines. The grid intervals are chosen depending on the scale of mapping and the number of observations per unit area desired. The nature of the terrain to some extent influences the grid interval chosen. In fairly simple, uniform landscapes the traverses may be fairly widely spaced whereas in complex terrains the traverses are closer together to ensure adequate coverage of all the variety of soil types. Generally the grid interval is determined by the size of the smallest mapping unit that can be shown on the map.

The square grid is commonly used, especially where the surveyor decides to use the national grid system shown on contour maps. In some cases a rectangular grid might be used such that the grid interval is smaller along one axis than on the other. The rectangular grid is useful if it is so orientated that the axis with the shorter interval coincides with the axis of maximum variation in soil and land types in the area. For example, where the structural trend is east–west, the grid is orientated in such a way that the shorter axis runs north–south to run down the hillslope profiles. However, at any given interval the grid which yields the largest number of sampling points is the triangular grid (see Fig. 5.1), in which the equidistant parallel lines are set at 60° from the vertical. Triangles are formed by drawing horizontal lines through the intersections (Peterson and Calvin, 1965). The triangular grid is the closest approximation to random sampling, the zig-zag journey from one point of the triangle to another being similar to a random walk over the terrain.

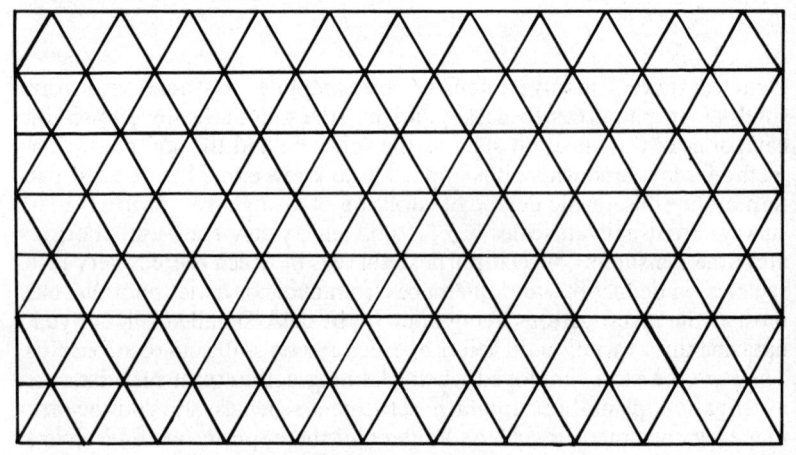

Fig. 5.1   A triangular grid

Grid mapping has fostered the development of quantitative soil survey (Rudeforth, 1969), in which soil profiles are examined at sites (grid points) chosen independently of natural or class boundaries. These provide a set of data points and the related information which may be analysed, manipulated, or grouped according to the needs of the various users. The grid points provide a means of determining mean values and the range of variability of soil properties in any map unit that may be delineated.

Free survey, or the mapping of soils by their boundaries, is the mapping procedure most commonly used by soil surveyors. Having established the mapping unit and the legend, the surveyor moves along lines where the field evidence suggests that one soil type gives way to another, occasionally inspecting soils on either side of each line to find the most accurate location of the boundary between the soil types. This technique presupposes that changes in soil conditions will be matched by corresponding changes in visible landscape features such as slope, drainage, vegetation, and so on. It is not thought to be a suitable method to use in large-scale or detailed mapping because many differences in soils which would necessitate division into separate units have no visible expression at the surface. Although mapping by the method results in accurately located soil boundaries, the number of observations may be insufficient to ascertain the uniformity of each map unit. The sampling density usually varies with each individual surveyor, and according to the complexity of the terrain, the distinctness of the soil boundaries, and the scale of mapping.

In physiographic mapping the terrain is classified into comparatively homogenous morphological units based on external properties of the soil or land as interpreted from aerial photographs or by special aerial sensors. The fieldwork is concerned primarily with identifying the soils in each unit and not with locating soil boundaries. This soil-mapping technique is the most economical of the three procedures in terms of time, money, and effort. The basis of this relatively new technique is the land system concept which is discussed later in this chapter.

3 Office and Laboratory Work

There are three major aspects of this stage of the soil survey all geared towards collating and processing the data on soils and other environmental factors collected before and during the field survey.

The first entails further photo-interpretation to make corrections on pre-field survey analysis and to fill in soil survey data collected during the fieldwork. In the United States, for example, the soil survey data are superimposed on controlled aerial mosaics. This produces uniformity of scales and makes it easy to match images between adjoining maps (USDA, 1966). Very commonly, these aerial mosaics are annotated and published as the soil maps themselves. But in many cases the information on the aerial photographs is transferred on to topographical base maps complete with key and symbols. This is the second aspect of the office and laboratory stage: the cartographic work. Most soil survey organizations have adopted the use of certain conventional symbols and colours for showing different groups of soils on maps. In western Nigeria, for example, the well-drained brownish soils are shown in different shades of orange and brown while poor to imperfectly drained soils are shown in blue and green. In Britain each genetic soil group is assigned a colour: e.g. poorly drained soils are shown in blue and green; brown earths in brown; podzols in red; and peaty soils in pink. The use of these conventional colours makes it easy to compare and match different soil maps and helps in the study of the relationship between the different soils and their environmental factors, especially slope or relief in general.

The cartographic work also includes the preparation of maps and illustrative diagrams that will accompany the soil survey memoir. Such maps and diagrams include those of the geology, the relief and drainage, climatic parameters (rainfall, temperature), hillslope models showing the toposequence of soils, soil profiles, and so on.

The third aspect is the analysis of soil samples in the laboratory. The nature and the amount of laboratory analytical work vary considerably between different soil surveys. In some the mapping and classification of soils are done entirely in the field without recourse to laboratory tests at all. But in detailed surveys standard tests on soils are carried out, namely, particle-size analysis, moisture content, calcium carbonate content, organic matter, nitrogen, cation exchange capacity, exchangeable cations, phosphorus, and free oxides, etc. In recent years the nature of the work has changed and has been expanded to include not only those routine tests on soil properties relevant to an understanding of soil–crop relationships but also fundamental micromorphological and clay- and sand-grade mineralogical studies. Most analytical tests are slow and laborious; they also contribute substantially to the total cost of a soil survey.

4 *The Soil Survey Memoir*
The final stage of the survey is the publication of the soil map and the soil survey report or memoir. The memoir usually contains (1) a general description of the environment, (2) an explanation of the system of soil classification and the soil mapping units, (3) an account of the field survey and mapping procedure, (4) the description of the distribution and spatial relationships of the major soil groups, (5) detailed descriptions of the morphological characteristics of the different soil classes, (6) analytical data on soil properties, and (7) a brief note on the use of the resource (e.g. land use, farm mechanization, the use of fertilizers, soil erosion, and conservation).

Soil Survey Interpretation

Soil survey interpretation is the analysis and evaluation of soils information for land-use and planning purposes. The aim is to show soil potentials and limitations for use; in other words, soil capability assessment. Soil is a very complex and variable resource; its suitability for various types of land use, arable farming, grazing, forestry, building, and road construction, etc. varies from place to place. Various interpretative soil maps can be produced showing the capabilities of the soils for various uses in any given area. Soil survey interpretation is undertaken not only by the pedologist or the agriculturist but also by the engineer and the hydrologist, the forester and wildlife conservationist, and the urban and industrial planner, to name a few. Such soil survey interpretations are encouraging landowners, planning officials, and various other interests to be more aware of the relationship between physical resources, economic considerations, and social needs.

In order to make the best use of an interpretative soil map the user must understand what soil properties have been used in the soil survey, the premises and criteria established for the interpretation of these properties, and the nature and variety of possible interpretations. For any given purpose, interpretation begins by an analysis of the present usages of the land to find out their relationship with the soil units. It is also to correlate characteristics common to soils with similar conditions, e.g. poor yields of a certain crop may be found to occur on all soils with poor drainage or soils deficient in certain chemical components. It is in this way that causal relationships are established between soil conditions and crop behaviour. The most important soil properties in terms of specific land-use types are also identified. However, it is not only the soil properties that are taken into consideration but also related factors of climate, relief, slope, ruggedness of the terrain, parent material, ground drainage, and vegetation. The socioeconomic base and the level of technology of the area in question should also be considered.

The commonest form of soil survey interpretation is that aimed at showing agricultural land capability. Every piece of land is assessed on its capability to sustain various types of agricultural activity. Land is classified into grades according to the most sustained use that can be made of it, as well as the extent to which it provides adequate protection from soil erosion and other forms of soil deterioration. The land capability classification system of the United States Soil Conservation Service is the most widely used system. In the system there are eight land capability classes. The intensity of land use decreases with increase in limitations and hazards to management practices. The limiting factors include slope of the land, the stoniness of the soil and ruggedness of the terrain, erosion, soil permeability and moisture-retentive capacity, soil depth, ground drainage and soil fertility, and the length of the growing season. Only soils in the first four classes are suitable for cultivation; the other land classes are used as pasture, woodland, or for wildlife conservation (see Table 5.2, pp. 95–6).

More recently, the Food and Agricultural Organization (FAO) of the United Nations carried out extensive soil surveys in Africa with a view to assessing the potential utility of the soils. In grouping the soils of

Nigeria into capability classes, for example, nine characteristics of the soils were identified as being crucial to their use for agricultural purposes in terms of present methods of farming. Each of the nine characteristics was designated by a letter: depth (P); texture and structure (T); base saturation (N); level of salinity (S); organic matter content (O); nature and amount of clay (A); mineral reserves in the soil (M); drainage (D); and moisture content (H) (FAO, 1965). The productivity of the soils is regarded as a function of these nine characteristics

$$P = f(P, T, N, S, O, A, M, D, H).$$

Each of the nine characteristics has a scale varying from zero to a hundred. Soils are graded according to their score on each scale. For example, with the first characteristic soil depth (P), soils are graded as follows:

Grade		Score
P_1	Soils with very little or no depth at all	5
P_2	Soils with a depth less than 30 cm	20
P_3	Soils with a depth between 30 and 60 cm	50
P_4	Moderately deep soils which have depths between 60 and 90 cm	80
P_5	Deep soils which have depths between 90 and 120 cm	100

In order to operationalize the equation, the factors are multiplied. The idea of multiplication is that limiting factors are made to over-rule advantages. For example, if a soil scores low on one characteristic and scores high on another, when they are multiplied the low value downgrades the high value.

Five productivity classes of soils are recognized in Nigeria based on the score of each soil group on all the nine characteristics. These are:

Grade		Score
1	Very high productivity	65–100
2	High productivity	35–64
3	Medium productivity	20–34
4	Low productivity	8–19
5	Very low productivity	0–7

The area covered by each grade of soil is given as follows:

Grade		Area in sq km	% of total area
1	Very high productivity 65–100	—	—
2	High productivity 35–64	50,400	5·52
3	Medium productivity 20–34	289,200	31·72

4 Low productivity 8–19	423,600	46·75
5 Very low productivity 0–7	148,800	16·31

Thus in terms of current agricultural practices only about 37 per cent of the total land area can be considered agriculturally productive. But there are a number of management practices which can be used to improve the soils.

The FAO system just described differs from that of the US Soil Conservation service on the grounds that the assessment is restricted to soil properties alone. It is therefore a soil and not a land capability assessment.

Soil in a Multidisciplinary Integrated Land Resources Survey

The discussion above should have indicated the crucial position occupied by soils in the natural environment. The important point about the natural environment is that all the components are inter-related. In other words, it is possible to talk of an 'environment system' and in recent years field scientists have adopted the integrated systems approach to the study of the environment. This conception of the environment as a system is not only of academic interest but is of the greatest value in planning the rational development and utilization of the earth's natural resources. This requires a knowledge of the mutual relationship between the elements as no one member of the system can be developed and exploited without taking cognizance of its role and position in the system and the effects which such a development would have on the whole system. Soil is therefore being studied more and more within the context of the total environment. This is the case, for example, in community ecology, biogeography, and landscape science. The idea of mapping soils by delimiting physiographic land units came from landscape science. Landscape science recognizes the possibility of dividing the landscape of any territory into a hierarchy of successively smaller units. These units can be grouped into recognizable patterns which recur in space. The same idea stimulated the attempts made by American geographers early in this century to divide the country into physiographic regions based on a common genesis. For example, Bowman (1914) divided America into physiographic provinces which were in turn subdivided into topographic types; while Fenneman (1916) also divided the country into physiographic provinces, sections, and districts. The advantages of these genetic landscape classifications became obvious to those concerned with land-use inventory surveys. These advantages included: (1) the close relationship between terrain and the distribution of human activity; (2) the relative homogeneity of the terrain types; and (3) with the development of aerial photography, the ease with which most land types could be recognized on aerial photographs. These landscape classifications remained the basis of land-use surveys in America until the 1950s.

The genetic systems of landscape classification, in spite of their usefulness, have many limitations, especially for purposes of defining precisely land-use and topography relationships to meet the requirements of present-day land-use planning and prediction. Hence the introduction of the concept of an 'integrated landscape' by the Commonwealth Scientific and Industrial Research Organization (CSIRO), Australia (Christian, 1958). This concept sees the region as deriving its character from a combination of inter-related factors such as climate, geology, relief, natural drainage, soils, and vegetation. Individual considerations of these environmental determinants in isolation from one another is of limited value; they must be studied from the point of view of their interaction with one another to appreciate fully the region as a whole. In characterizing and classifying the landscape one should look at the landscape itself rather than the genetic factors. The relative importance of each factor varies from place to place, thereby leading to differentiation in landscape patterns. But the number of landscape patterns (or regions) that can be found in any one country is not unlimited; they are often found in recurrent patterns over the earth's surface.

A region in which there is a recurrent pattern of topography, soils, and vegetation has been termed a 'land system' by the CSIRO. This property of recurrence means that knowledge of one area can be transferred, through a system of analogies, to another area with similar environmental characteristics (Areola, 1971, 1974). The land system is a convenient mapping and classification unit on a fairly small scale. It can be subdivided into smaller units, which are more uniform in character, called 'land facets'. The land facet is a fundamental terrain unit which has been defined as 'a recurrent terrain feature sufficiently small to be uniform for all practical purposes—or, if not uniform, showing variations in the same sense wherever it occurs' (Beckett and Webster, 1965). The idea of the land facet is reflected in Moss's ecological concept of the 'habitat-site', a unit of land with a characteristic slope and form which constitutes a distinct habitat for plant growth (Moss, 1968). The land facet is sufficiently homogeneous to be used as the basis for drawing inferences about a wide range of terrain attributes including soils.

Land facets used as soil-mapping units have been termed 'pedomorphological units'. The use of land facets in soil mapping derives from the fact that such units have the important properties of position, shape, and area. They are, therefore, easy to recognize in the field, unlike the traditional soil-mapping unit—the soil series—which is a conceptual rather than a real unit. Furthermore, in terms of land-use planning, such units which express the totality of environmental conditions are considered more useful as planning units than the soil series. It is important to point out, however, that, in practice, land facet mapping has been beset with the same problems as soil series mapping and it has been much criticized for this (Moss, 1969; Thomas, 1969).

Landscape classification is a prelude or first step to land evaluation which is the assessment of the suitability of land for use in agriculture, forestry, engineering, hydrology, regional planning, recreation, etc. (Stewart, 1968). It involves not only a consideration of the natural resources but also such human resources as technology, labour, and finance. Land evaluation is therefore the economic appraisal of land resources given a certain level of technology to cater for the needs of the community. The work involves the joint effort of various disciplines and field scientists, an approach which ensures complete coverage of all aspects of both the natural and the cultural environment.

Soil Survey with Special Reference to Africa

The term 'difficult country survey' has been used to describe soil surveys in Africa. Some of the reasons for this rather harsh description can be deduced from the following extract from Smyth and Montgomery's account of the soil survey of central western Nigeria:

Soil survey work on an extensive scale is exceptionally difficult in the rainforest zone of western Nigeria. The natural vegetation is dense and the road network, although greatly improved during the period of the survey, is often inadequate. Such topographical maps as exist are on a scale of 1 : 125,000 and are often too inaccurate to be used as base maps, while geological maps for only two small areas were available at the time of the survey. Little use could be made of air photographs since, for climatic reasons, the quality of the available photographs was extremely poor. Finally, over the metamorphic rocks, the soil pattern is very complex and a large and systematic sampling programme is required to obtain a representative sample of the soils.

Smyth and Montgomery, 1962, p. 21.

There are also problems of organization, finance, and technology. Very few (if any) survey units in Africa have reached the same level of organization and efficiency as those of the more established soil surveys in the advanced countries of Europe and America. Much soil survey work in Africa has been defeated by the sheer lack of trained personnel and professional men and by the magnitude of the work to be done as many countries stretch over wide expanses of land. In many countries the soil survey has been rather haphazard with little planning, co-ordination, or standardization of the mapping procedure. Many of the methods in use are adaptations of those employed overseas and, often, expatriates are heavily relied upon for their proper interpretation and application to local conditions. The resources available to the soil survey units, in terms of money and equipment, are extremely limited because of the very poor socioeconomic environment in which they operate. Transport and labour often account for much of the total cost of a soil survey. Finally, whereas in the advanced countries the field survey team consists of not more than two or three persons, in the difficult environment of most developing countries soil survey requires more labour. For example in the soil survey of central western Nigeria, a typical field team comprised:

1 soil scientist (or soil survey officer);
1 or 2 supervisory assistants (agricultural assistant grade);
10–15 junior assistants—who actually did the soil sampling;
50–60 labourers—engaged in cutting traverses through dense forests and thickets, in swampy valleys and over rugged hills;
2–3 lorry drivers.

Soil survey work in Africa began in earnest after the Second World War, although before then exploratory field investigations of soils had been made sporadically in some areas in the 1920s and 1930s. In West Africa, for example, soil investigations were made by agricultural chemists under the colonial administration, while in East Africa the work of G. Milne deserves special mention, mainly on account of the 'catena concept' which he formulated (Milne, 1935).

The recognition of a need for soil maps and the establishment of soil surveys in African countries was brought about in part by the failure of agricultural development schemes after the Second World War (e.g. the groundnut schemes in Nigeria and Tanganyika). The decision in 1953 to compile a soils map of Africa at a scale of 1 : 5,000,000 to aid the planning of agricultural development also encouraged soil survey work because each country was required to submit a national soils map. The project, undertaken by the Commission for Technical Co-operation in Africa (CCTA) and completed in 1964, gave a broad description of soil distribution in Africa. This study is discussed at length in Chapter 7. The project has value in providing a broad picture of the soil resources of Africa. But, like most soil surveys in Africa, the soil investigations were of an exploratory kind; there is an increasing demand for more detailed soil surveys to meet present-day land-use planning requirements. This type of work is now being done by soil survey units which are usually departments within the Ministries of Agriculture and Natural Resources in the different countries. However, most of the available surveys are still of the reconnaissance and semi-detailed types, detailed surveys being largely restricted to project areas, e.g. the sites for farm settlements, dam, and irrigation projects.

In addition, the Land Research Division (LRD) of the British Ministry of Overseas Development has conducted soil surveys in many African countries (Nigeria, Lesotho, Botswana, Kenya, and Tanzania) on a contract basis. In Nigeria the LRD has completed a land resources survey of about 62,000 square kilometres of land in Adamawa and Sardauna Provinces (now Gongola State), as well as a soil survey of the savanna lands of northern Oyo Province in Oyo State. The LRD surveys involve mapping and classifying the terrain units; describing, among other things, the soil attributes, and attempting to assess the suitability of the land for various types of crop, and for forestry and grazing (see Bawden and Tuley, 1966). The Institute of Agricultural Research (IAR), Samaru, Zaria, is another important group involved in soil surveys in Nigeria. We shall describe a few examples.

The Soil Survey of Western Nigeria (now broken up into Bendel, Oyo, Ogun, and Ondo States) began systematic field investigations of soils in 1951 under the direction of H. Vine. The survey method adopted, called the traverse method, was largely based on that employed by C. F. Charter in Ghana. In this method, soil samples are collected at regularly spaced sampling holes along parallel compass traverses. The density of sampling is controlled by the spacing of the traverses and of the sampling holes along them. On the basis of the density of sampling, three types of soil survey were made: (1) reconnaissance survey, in which the traverses were spaced 6·5 km apart and the sampling holes 200 m apart; (2) intensive reconnaissance, in which the traverses were spaced 1·6 km apart and the sampling holes 100 or 200 m apart; and (3) semi-detailed survey with the traverses and the sampling holes 100 and 200 m apart respectively.

The traverse method, though expensive, slow, and laborious has been found to be well-suited for soil survey in difficult environments. The advantages claimed for the technique are:

1 In an area where motorable roads are few and far between and where the natural vegetation is thick, the traverses provide a means of access.
2 In the absence of adequate base maps, the regular spacing of the sampling holes along compass traverses makes it possible to locate and map the holes accurately.
3 The traverses can be orientated in such a way as to cross the topographic grain of the country at right angles to ensure adequate coverage of the variety of landscape types and altitudinal zones in the area.
4 The fieldwork, consisting of traverse cutting, soil sampling, and land-use recording, is simple enough for junior assistants to carry out, after only a short period of training. This leaves the few trained professional staff free for the more demanding aspects, such as the recognition and measurement of soil properties, which require greater skill and knowledge.

5 The grid traverse provides a uniform sampling pattern and makes the statistical handling of data possible, for example, to estimate the proportion of each soil type or land-use class or the relationships between them in the area.

The terms and definitions used in the soil descriptions were carefully worked out to suit local conditions. An attempt was made to correlate these definitions with those in use elsewhere as, for example, in the textural classification of the soils (Table 5.1).

TABLE 5.1

Comparison of Textural Classes, Western Nigeria: USDA

Western Nigeria Soil Textural Classes		USDA Soil Textural Classes
Class	*Approx. % of silt + clay*	*Class*
Sand	less than 8	Sand
Slightly clayey sand	8–15	Loamy sand or sand
Clayey sand	15–25	Mainly sandy loam
Very clayey sand	25–35	Mainly sandy clay loam
Sandy clay	35–55	Sandy clay (+ some sandy clay loam)
Clay	more than 55	Clay

Compiled from Smyth and Montgomery, 1962, p. 37.

The soil survey section of the IAR, Samaru, Zaria, has carried out extensive surveys in the northern parts of Nigeria mostly at the reconnaissance and semi-detailed levels. The object in all cases was to describe, classify, and map the soils and so provide the basic information necessary for proper land-use planning. There is no systematic soils sampling system as in western Nigeria. Rather most of the fieldwork to investigate soils was done on 'treks' along village roads and the several footpaths which are the main arteries of communication in the rural areas of the north. The roads and footpaths are 'levelled and augered' at intervals and pits are dug at representative sites. Most of the soil boundaries are located by aerial photo-interpretation. This is made easy by the close relationship between soils on the one hand and the rocks and relief on the other (see *IAR Soil Survey Bulletin,* Nos 31, 34, 36).

In the French-speaking areas of Africa soil survey work of a more extensive and exploratory nature has also progressed. The work of the

Office de la Recherche Scientifique et Technique Outre Mer (ORSTOM) of Paris is worth mentioning. This is because most of the French areas in which the organization has worked cover the savanna and more arid parts of Africa. It is therefore largely responsible for much of our knowledge of the soils of these areas. In their genetically based worldwide soil classification scheme—the ORSTOM system—tropical soils are given prominence.

With regard to the classification of African soils, the work of the Belgian Institut National pour L'Etude Agronomique au Congo (INEAC) is also important. The INEAC classification system (see Chapter 7) was specifically designed for intertropical soils (D'Hoore, 1968). A lot of the research on laterite and other ferruginous tropical soils has been carried out in the dry French-speaking areas.

In conclusion, one can say that much has been achieved in soil surveys in Africa since the end of the Second World War, but there is still a great deal of ground to cover. Most of the surveys so far have been exploratory in nature in which case the soil classes are only broadly defined. There is an urgent need for more detailed and fundamental soil investigations to meet present-day problems, e.g. those related to farm mechanization, liming, and the use of fertilizers.

TABLE 5.2

The USDA Land Capability Classification System

Class	Limitations	Capability
I	Few	Deep, well-drained fertile soils with high water-holding capacity; need only ordinary crop management practices to maintain their productivity; can be cropped intensively, or used for pasture range, woodland, or wildlife
II	Some limitations that reduce choice of plants including one or more of: gentle slopes; moderate erosion hazards; inadequate soil depth, structure, and workability; slight drainage impedance; soil salinity	Similar to Class I but can sustain less intensive cropping systems except with some conservation practices: e.g. terracing; strip cropping; green manuring; etc.

TABLE 5.2 (*contd*)

Class	Limitations	Capability
III	Severe limitations that reduce choice of plants: e.g. moderately steep slopes; high erosion hazards; very slow permeability; shallow depth; low water-holding capacity; low soil fertility; soil salinity; etc.	Require special conservation practices to grow same crops as on Classes I and II land
IV	Very severe limitations on the choice of plants: e.g. steep slopes; severe erosion; shallow soils; low water-holding capacity; poor drainage; severe salinity	Has more limited alternative uses than Class III land; very careful management may be required; crops that will provide effective cover for the soil are required
V	Very severe limitations to cultivation: e.g. severe erosion hazards; frequent inundation; short growing season; stony or rocky soils	Unsuited to cultivation but good enough for improved pasture, range, woodland, wildlife
VI	Very severe limitations as for IV but more profound	Use restricted to pasture or range, woodland or wildlife
VII	More severe limitations than in VI	Pasture improvement is impractical; may be used for rough pasture, woodland, or wildlife
VIII	So severe that the land should not be used for any kind of commercial plant production	Use restricted to recreation, wildlife, water supply, or aesthetic purposes

Compiled from Buckman and Brady, 1966.

Chapter 6

METHODS OF STUDYING THE SOIL PROFILE IN THE FIELD

The unit of soil study is the two-dimensional cross-section known as the soil profile. The soil profile provides a record of past and present evolutionary factors and processes in the development of the soil which can be analysed and described. In order to gain a full appreciation of its nature the soil must be studied in perspective in its natural environment, that is, in the field. The study of the soil in the field assumes especial importance when dealing with soils at the local level where it is possible to study them in relation to the environmental factors and land use.

Some of the soil properties—mostly the chemical properties—are not visible and so cannot be observed in the field. Such properties are best analysed in the laboratory (see Appendix). Other properties, however, are observable or can easily be inferred from the visible features of the land; they can be measured easily and quickly, with a fair amount of accuracy, directly in the field. In this chapter, we describe in some detail methods of, and the procedure for, examining, measuring, and describing observable soil morphological characteristics in the field. The USDA *Soil Survey Manual* and Clarke's *The Study of the Soil in the Field* provide much of the material in the chapter, supported by the authors' own experiences in the field.

Soil Survey Equipment

There is no standard list of equipment for use in the study of soils in the field; the number and type of items vary depending on the scale of the survey and the soil properties being investigated. The most commonly used ones fall into the following categories:

1 *Base map* This is usually a good topographical map of suitable scale; air photographs are also widely used for similar purposes. During the survey, points at which soils have been examined are located on the base map or photograph, soil site characteristics are recorded, and soil boundaries marked.

2 *Equipment for taking topographic data* such as altitude, slope angle, and aspect, etc. These include the altimeter or aneroid, the abney level, clinometer or theodolite, the prismatic compass, and ranging poles, i.e. common survey instruments (for description see Ajaegbu and Faniran, 1973).

3 *Equipment for digging and preparing the soil profile pit* This

includes the spade or shovel, pick-axe, hand trowel or a sharp curved knife, and a brush. The hand trowel or knife is used in smoothing the face of the profile and the brush in removing fine soil particles which might contaminate other horizons. Where the aim is just to inspect soils without actually digging pits the equipment needed is an auger or a corer. The T-shaped screw auger is the commonest. It is twisted into the soil to the required depth and then pulled out to study changes in soil properties with depth from the soil sticking to the screw. Unfortunately, deeper samples are liable to contamination as the auger is pulled out of the soil through the top layers, while, in the process of twisting, soil structure is disrupted. A corer (or borer) is better in this regard, as it enables cylindrical columns of soil to be gouged out without greatly disturbing soil structure and constitution. It is most useful in moist and fairly firm soils, since it cannot hold the particles together in loose and dry soils.

4 *Equipment for making linear measurements* This includes a tape (the steel tape can usually stand rough field conditions more than the cloth type), a foot-rule, and a graduated measuring rod.

5 *The soil-testing kit* This is a collection of chemicals and portable instruments for making simple tests on some soil chemical properties such as pH and carbonate content, etc. It includes: either a portable, battery-operated pH meter or a soil-testing indicator with its accompanying colour chart; distilled water; suitable containers (beaker, test tube, porcelain tile); 10 per cent solution hydrochloric acid; gypsum blocks and recorder for measuring electrical conductivity (moisture content); etc.

6 *The Munsell colour chart* For determining soil colour.

7 *Equipment for collecting soil samples* For collecting small samples from individual soil horizons, polythene bags are commonly used. They preserve soil moisture and prevent the sample from drying out and from being oxidized and decomposed. However, it is thought that, with time, some of the soil chemical substances, especially phosphorus compounds, are gradually changed by biological action. Cartons and other types of container (e.g. brown envelopes) can be used but samples are more likely to dry out. For cutting soil monoliths, special wooden or steel containers are needed. These are fully described later in this chapter.

8 *Sundry items* The field pedologist would take along to the field writing material, raincoat, boots, and a good camera. Nowadays colour camera films are generally available and they are invaluable as a means of making a record of the soil profile as seen in the field.

Stages in the Study of a Soil Profile in the Field

The organization of the fieldwork will vary from place to place and with individual surveyors. It is possible for a single surveyor to do all the work, if he so chooses. But for speed and efficiency the field team should consist

METHODS OF STUDYING THE SOIL PROFILE IN THE FIELD

of at least two—the surveyor and an assistant; a third man to take on the job of digging and preparing the soil pits is also essential. While the surveyor is making his measurements and analyses, his assistant records the data and other information passed on to him by the surveyor. The assistant also helps with the slope and compass-bearing measurements and in carrying equipment from one site to another. The surveyor chooses his lines of transect carefully so that he is also able to gain a panoramic view of the entire landscape as he follows his course.

The aim of direct field survey is to gain first-hand knowledge of the soil in its natural environment; but the realization of this objective can be ruined if the field course is not properly planned and systematized. This is why it is necessary to proceed in definite, laid-out stages as shown on field sheets or proformas, which are plans showing the order in which soil site and profile characteristics are to be examined and described in the field (Table 6.1a and b).

TABLE 6.1 *Proforma*
(a) Soil Site Characteristics

Locality	Site no.	Grid reference
Soil series	Major soil group/catena/land system	
Topographic position	Gradient	Aspect
Slope form	Micro-relief	
Site drainage		

Parent material
Solid rock
 	
Superficial deposits:
  Mode of formation
  Lithology
  Structure
	Vegetation and land use	

Soil Site Characteristics
The basis of the field study of soil profiles derives from Dokuchaev's doctrine that the soils found in any area are a function of environmental factors such as the climate of the locality, the nature of the parent material, the character of the vegetation, the age of the country, and the relief of

(b) Soil Profile Characteristics

Horizon	Thickness (cm)	Boundary	Colour	Textural class

	Horizon:	1	2	3	4	5

Stoniness
  Abundance
  Class (size)
  Mineralogy						

Structure:
  Type
  Size
  Consistence						

Drainage:
  Class
  Depth of gleying
  Abundance of mottles
  Mottle colours						

Organic matter:
  Litter layer thickness
  Composition of litter
  Root abundance
  Root size
  Type of humus						

Horizon	pH	$CaCO_3$	Secondary minerals

Sketch:

the locality (see Chapter 1). Therefore, when the surveyor has chosen the point in the field where he intends to examine the soil profile, his first assignment is to study and describe the site characteristics, possibly in the following order:

LOCATION AND IDENTIFICATION

At the head of the field sheet the index number of the site is recorded together with the soil series name and its subdivisions (soil type, soil phase). The major soil group to which the series belongs may also be entered on the sheet. At times, as in East Africa, the name of the catena or association, or, as in Australia, the land system to which a soil belongs is recorded in place of the major soil group.

The site is then located on the map and is usually given the name of the locality in which it is situated. The map reference can be given in terms of longitude and latitude when the area covered is large or where there is no system of grid references. In small geographic areas points are more accurately located by the use of grid references.

TOPOGRAPHIC DATA

The topographic parameters often recorded include altitude, slope angle, shape and aspect, and micro-relief. The altitude of the site can be derived from contour maps by interpolating between contour lines and spot heights. But such measurements can only be estimates; the smaller the contour intervals, the better the estimate. More accurate readings of altitude are possible with an altimeter or the field sketching aneroid. These instruments, however, need to be used with great caution.

Slope angle is measured and expressed commonly either as a percentage or as a degree slope. It is also possible to express slope as a ratio of vertical distance (height) to the equivalent horizontal distance . Table 6.2 shows the relationship between the three different methods of expressing slope.

Slope form is as important as slope angle in understanding the movement and distribution of weathered material on the hillslope. Slope forms can be divided into two broad categories: uniform or even slopes, also called rectilinear slopes; and complex slopes. Complex slopes consist of various sections with varying shapes: concave, convex, terrace, step, and flat. The hillslope can be divided into sections known as morphological units which are areas of slope with uniform and distinctive shapes, angles, and direction. Fig. 6.1 shows a hillslope model and the possible classification of the morphological units.

Micro-relief refers to minor variations in the surface morphology of the macro-slope units. Thus a slope can be described as rolling, gently undulating, rugged, rocky, or broken (e.g. dissected by rills, gullies, and erosion scars).

Aspect is the direction in which a slope is facing and is usually measured as the direction of a compass bearing taken at right angles to the slope. This direction can be expressed in degrees (departures from the magnetic north); by the cardinal points (N, NE, etc.); or by both. It is not possible to determine the aspect of flat, nearly flat, or rolling surfaces; so that aspect can only be measured where there is an appreciable slope of at least 3°.

TABLE 6.2
Measurement of Slopes: Table of Equivalents

Angle of inclination	Equivalent tangential ratio	% slope m per 100 m of distance
0° 10′	1 in 343	0·29
0° 20′	172	0·58
0° 30′	114	0·87
1° 0′	57	1·74
2° 0′	28	3·49
3° 0′	19	5·24
4° 0′	14	6·99
5° 0′	11	8·75
6° 0′	10	10·51
7° 0′	8	12·28
8° 0′	7	14·05
9° 0′	6	15·84
10° 0′	6	17·63
11° 0′	5	19·44
12° 0′	5	21·26
13° 0′	4	23·09
14° 0′	4	24·93
15° 0′	4	26·80
16° 0′	3	28·67
17° 0′	3	30·57
18° 0′	3	32·49
19° 0′	3	34·43
20° 0′	3	36·40

From G. R. Clarke (1957).

Based on the angle of inclination, slopes may be classified as follows:

0–½°	Flat	6½–13°	Strongly sloping
1–2½°	Gently sloping	13½–19°	Moderately steep
3–6°	Moderately sloping	19½–31°	Steep
		31° and over	Very steep (cliffs)

After Curtis, Doornkamp, and Gregory (1965).

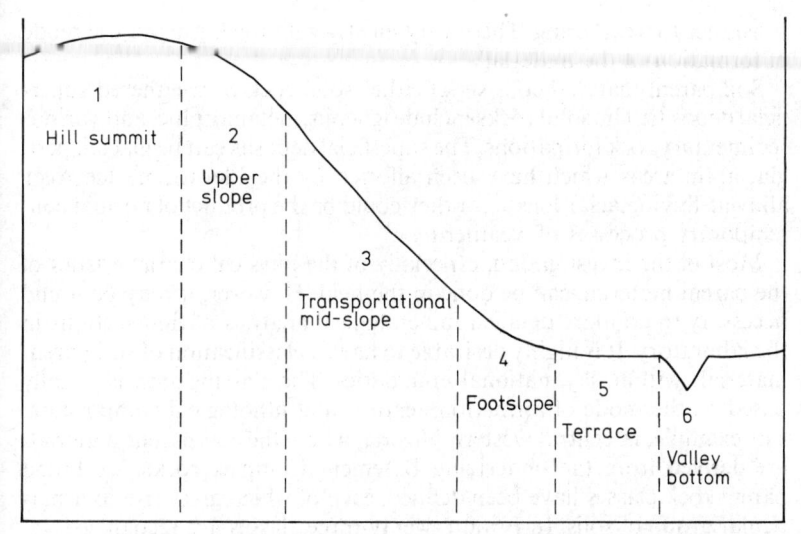

METHODS OF STUDYING THE SOIL PROFILE IN THE FIELD

Fig. 6.1   A simple hillslope model showing morphological divisions

(1 Hill summit, 2 Upper slope, 3 Transportational mid-slope, 4 Footslope, 5 Terrace, 6 Valley bottom)

SITE DRAINAGE

After describing topography, the drainage of the site (as opposed to soil drainage) comes up next for study. This includes, in the first place, a consideration of the main drainage lines such as streams, springs, rills, and gullies. The density of drainage lines is an index of the degree of dissection of the landscape and this is important in the erosion and deposition of material on the land surface. The width and shape of the valleys also can give an idea of the maturity of the streams and of the landscape. Secondly, the drainage condition of the ground, which is often related to the depth of the water table and the textural and lithological characteristics of the parent materials, is considered. Thus, a soil site may be described as well drained, poorly or imperfectly drained, marshy or waterlogged, liable to flood, and so on. Finally, artificial drainage lines, such as ditches and drainage pipes, should also be taken account of. Standing water on the surface like pools, ponds and lakes, also form a component part of the soil site drainage.

PARENT MATERIAL

The soil parent material holds the clue to many soil properties and should therefore be studied in detail. The most significant aspects of the parent material which may have important consequences for the soil are the mineralogy, texture, degree of cohesion, drainage characteristics, and resistance to weathering. These vary mostly with the form and the mode of formation of the material.

Soil parent material consists of either solid rock or weathered superficial deposits. The solid rocks include igneous, metamorphic, and various sedimentary rock formations. The superficial deposits can be glacial, periglacial (in areas which have been affected by the Pleistocene Ice Age), alluvial, fluvioglacial, loessic, or they could be the product of normal contemporary processes of weathering.

Most of the investigation, especially of the physical characteristics of the parent material, can be done in the field. However, it may be found necessary to do more detailed mineralogical analysis of thin sections in the laboratory. It is highly desirable to have a classification of soil parent materials within, say, national boundaries. The classification is usually based on the mode of formation, texture, and lithological composition. For example, in central western Nigeria, where the soil parent materials are derived from the underlying Basement Complex rocks, six broad parent rock classes have been defined, each of which gives rise to a particular group of soils. In Britain twenty-three classes are recognized, including various primary and sedimentary rock formations, glacial, alluvial, and organic (peat) deposits.

VEGETATION

Vegetation is not only a dominant factor influencing soil formation and determining its nature, it is also a reliable indicator of soil conditions, especially soil drainage, soil reaction, and nutrient status. The study of the soil site plant cover involves a consideration of (1) the various plant species present, (2) the vegetation structure, (3) the density of cover of the various strata of plants, and (4) the nature, composition, and amount of litter on the soil surface.

The most taxing part of the study of the vegetation could be the identification of plant species. But this can be facilitated by using handbooks which describe the distinctive features of the different plant species, illustrated with photographs and drawings.

LAND USE

The description of the vegetation and of land use always go hand in hand. Broad land-use categories such as forest, grassland, farmland, etc. are listed. If it is farmland, the various crops grown are listed; the size, shape, and orientation of farm plots are also expressed. In grassland used for grazing the animals under pasture should be listed. Farming practices such as terrace farming, contour ploughing, clay subsoiling in areas of poor drainage, and artificial drainage, which are common in certain parts of the world, or bush burning and soil heaping, commonly practised in the tropics, should also be mentioned if observable in the field.

METHODS OF STUDYING THE SOIL PROFILE IN THE FIELD

CLIMATE

Climate cannot be observed or studied in the field at the time of the soil survey. Data on the climatic elements have to be collected from the appropriate sources, usually from the periodic tables of the meteorological office or of local schools and colleges. If possible data should be collected on the following: (1) annual rainfall, its monthly and seasonal distribution, and incidence of drought; (2) mean minimum and mean maximum temperatures, extremes of temperature, and distribution of frost periods (if in high latitudes); (3) snowfall and snowcover; (4) relative humidity; and (5) wind direction and the nature of prevailing winds.

Soil Profile Description

PREPARATION OF THE PROFILE PIT

The soil site having been described, the soil profile pit can then be dug and prepared for detailed investigation. On the making of the profile pit, Clarke gives the following instructions:

> The pit should be rectangular in plan, large enough for a man to sit comfortably in the bottom, and as deep as is necessary to expose the parent material. The position when possible should be oriented in such a way that the light shines into the end of the pit and indirectly illuminates both long sides at the same time. When digging pits into hillsides it is particularly necessary to prepare the *sides* running down hill for the profile records. The 'cut-off' into the hill gives a false impression of the horizons by portraying them as approximately horizontal whereas in fact the true horizons will slope more or less as the slope of the hill.
>
> G. R. Clarke, 1957, pp. 51–2

The face of the profile can be smoothed by the use of a small curved knife or a hand trowel. It is recommended that the soil face be allowed to dry a little before commencing detailed investigation to allow some of the more important structural features to show up clearly. It is very important, too, that particles from one soil layer are not allowed to contaminate others; hence a small hairbrush should be near at hand to remove such offending particles.

Soil profile description may proceed as shown below:

1 Identification and naming of soil layers or horizons.
2 Description of soil colour and the Munsell notations.
3 Texture.
4 Coarse fragments (stones): shape, size, abundance, and distribution.
5 Structure and constitution.
6 Soil drainage (moisture content, gleying, and mottling).

7 Organic matter (nature and distribution, including plant roots, and faunal influence).
8 pH.
9 Carbonate content.
10 Secondary chemical deposits.

SOIL HORIZONS
The classification and nomenclature of soil horizons are based on their genetic characteristics. The division can usually be determined in the field by noting variations, down the soil profile, in such visible features as colour, structure and degree of compaction, moisture condition, size and frequency of stones, distribution of organic matter, and plant roots, etc. The horizons are named, from top to bottom, by the letters A, B, C, and other conventional symbols such as numerals, e.g. A_1, A_2, A_3 for divisions of the same genetic horizons; brackets ( ) for a not-fully-developed or transitional horizon; and lower-case letters used as suffixes to denote special characteristics, e.g. Ah for humic, Ae for eroded, Ap for ploughed agricultural layer, and Bs for layers rich in sesquioxides.

It is not always possible to name the horizons in the field as this might require careful consideration and assessment of the sum total of horizon characteristics including those analysed in the laboratory. Hence, the horizons are just numbered serially, Layer I, Layer II, etc. In many tropical soils, for example, the layers cannot always be named by the normal conventional symbols A, B, C, due to the length of time pedogenesis has been taking place and the tendency towards homogenization. Horizon differentiation below the top humus layer is weak in many soils, while some soil profiles are polycyclic and the layers can be related to past rather than contemporary pedogenic processes.

The thickness of each horizon identified is then measured to give an idea of the regularity of each layer from place to place. Some soil horizons form a continuous layer, some are disjointed, while still others are found in pockets within other layers.

Two aspects of horizon boundaries are also worth noting: their clarity and regularity. The clarity with which the boundaries between layers are defined may be described as follows:

Sharp: the horizon boundary occurs over a very short distance, not more than 2·5 centimetres.
Clear: the horizon boundary occurs over a greater distance, but is not more than 5 centimetres.
Merging or Gradational: the horizon boundary occurs over a distance of a few centimetres.

The regularity with which the boundaries run horizontally may also be described as follows:

Smooth or Even: the boundary runs parallel to the land surface.
Irregular: the boundary is neither straight nor parallel to the land surface. Irregular boundaries can be wavy, in smooth wave-like curves; or angular.

COLOUR

Soil colour should be determined in bright light, for the intensity of light falling on the sample affects the shade or brilliance of the colour. The colour of a dry soil differs slightly from that of a moist one, hence it is necessary to specify at what state, dry or moist, the colour has been read. It is generally advisable to measure both the 'dry' and the 'moist' soil colour. The Munsell colour chart is widely used for determining soil colour. But the old descriptive terms like orange, red, yellowish-brown, chestnut, etc. are still useful for they portray to the user some picture which cannot easily be appreciated using the Munsell colour notations.

The Munsell colour chart and its use The Munsell colour system describes colour in terms of three variables, *hue, value,* and *chroma.* Hue is the dominant spectral (or rainbow) colour, e.g. red, yellow, yellow-red, etc. It is a function of the wavelength of light reflected by the soil. Value describes the variation of colour between absolute white and absolute black, that is, its apparent lightness. It is thus a function of the intensity of light. Chroma refers to the purity of the hue or its departure from neutral (grey) of that same colour.

In the Munsell system, hue is regarded as varying along a continuum ranging from red to green. This continuum is divided into the major spectral colours, red (R), yellow-red (YR), yellow (Y), green (G), etc. The range of each major colour is divided into 10, e.g. the red (R) varies from 0 to 10R. The arrangement is such that the highest value in one grade corresponds with the zero-point of the next higher grade (being on a continuous scale). Value is also subdivided into 10 ranging from 0 for absolute black to 10 for absolute white. In like manner chroma has 10 subunits; pure grey has a chroma of 0; increasing departure from grey is indicated by serial numbers 1–10. In the chart there is a page for each grade of hue (but in the field handbook only those at 2·5, 5·0, 7·5, and 10 are shown). On each page the various grades and shades of colour are represented by colour chips and arranged in order. The value scale is shown along the vertical axis and chroma along the horizontal.

To determine soil colour by the Munsell colour chart the procedure is as follows:

1 Take a small sample of soil. If desired the soil is moistened simply with saliva.

2   The colour of the soil is then compared with each page in the Munsell handbook to find the closest match, i.e. the hue which contains standards (colour chips) closely resembling the colour of the soil.

With experience, it does not take much time to locate the hue of a soil sample; usually as soon as an experienced soil surveyor looks at a sample of soil he has some idea of its hue and where to look in the chart to find the closest colour match.

3   The soil is then compared in turn with each of the colour chips to find the closest match. The value is read along the vertical and chroma along the horizontal axis.

The colour notations are recorded thus, 10YR 5/8, in which 10YR is the hue, value = 5 and chroma = 8. It is possible that the soil colour does not match any one of the colour chips very well but lies between two chips. Since the colour notations are supposed to be on a continuous scale, fractions can be found between the grades of value and chroma. Thus the colour of a sample may be designated 10YR 5·6/8 or 10YR 5/7·4 as the case may be. The colour name is usually written below the colour chip, e.g. 10YR 5/8 is yellowish-brown.

It is possible to find a soil horizon with more than one colour. This is especially the case in a gleyed and mottled soil. In such a case the Munsell notations of all the colours present, including the mottles, should be determined and recorded.

SOIL TEXTURE

Texture is defined as the relative amounts of sand, silt, and clay in the soil. The actual amounts can only be obtained by fairly elaborate laboratory techniques. Soil texture determination in the field involves merely the assignment of the soil to a textural class based on the 'feel' of the soil mass when moistened and rubbed between the fingers. Sand has a gritty feel and the individual grains can normally be seen with the naked eye. Silt has a rather smooth silky or soapy feel, while clay is rather tough and sticky.

Another simple test which is carried out to estimate textural class is to take a small quantity of soil, moisten it, and try to roll it into a thread or mould it into a ball. For example, if the soil can be rolled easily into a thread it is probably clay-loam, but if it just makes a thread it is likely to be loam. If the thread can be bent easily or moulded into a ball, the soil is probably clay. There are twelve textural classes in all according to the relative proportions of sand, silt, and clay in the soil. These textural classes and their diagnosis in the field are listed in Table 6.3.

TABLE 6.3
Soil Textural Classes: Definition and Identification

	Soil textural class	Definition and identification
1	Sand	Soil consisting mostly of sand ($>85\%$) and very little clay ($<5\%$). It is loose and will not cohere or roll into a thread.
2	Loamy sand	Soil consisting mostly of sand (c. 70%) but with sufficient amount of clay (c. $10–20\%$) to give it some cohesion; but it will not roll into a thread.
3	Sandy loam	Sand still forms a large proportion ($>50\%$) but there is sufficient clay (c. 20%) to allow the soil to cohere and mould fairly easily when moist without being sticky. It will not roll into a thread.
4	Loam	Sand is less conspicuous (c. $40–50\%$) and clay and silt form an almost equal proportion. The soil coheres easily and is slightly sticky when moist; but it will not roll into a thread.
5	Clay-loam	The soil contains a substantial amount of clay ($>30\%$) to mask the presence of sand ($20–40\%$). It is sticky when moist and although it will roll into a thread, the thread will crack if bent to form a ring.
6	Sandy clay-loam	The sand content is more obvious ($>45\%$) giving the soil a gritty feel in spite of the substantial amount of clay (c. $20–35\%$). It will roll into a thread but the thread cannot be bent to form a ring without cracking. It is sticky when moist.
7	Sandy clay	Clay ($>35\%$) and sand ($>45\%$) are dominant. Though sticky and plastic when moist, it has a gritty feel. It will roll into a thread but the thread cannot be bent to form a ring without cracking.
8	Clay	Soil in which the clay accounts for more than 40% of the mineral fraction. It is very tough and sticky when moist. It is easy to roll, bend, and mould into a ball.

TABLE 6.3 (contd)

	Soil textural class	Definition and identification
9	Silt	Soil consisting mostly of silt ($>90\%$). It has a smooth soapy feel. It is friable and coherent. It will not roll into a thread.
10	Silty loam	Soil consisting mostly of silt ($>70\%$); and some sand ($<25\%$). It is friable and coherent but it will not roll into a thread.
11	Silty clay-loam	Soil containing a large clay content ($c.$ 20%) but with a fair amount of sand ($c.$ 20%) and sufficient silt ($c.$ 60%) to give it a smooth soapy feel, though slightly sticky. It will roll into a thread but the thread cannot be bent to form a ring without cracking.
12	Silty clay	Silt ($c.$ 60%) and clay (40% or less) are dominant. It is slightly sticky, with a smooth soapy feel. It will roll into a thread but will crack if bent to form a ring.

From USDA (1951).

The determination of soil textural class in the field requires much skill and experience. The accuracy with which it is done varies with the individual.

Sometimes the soil surveyor may adopt another method. He may estimate the proportions of only two of the textural grades, say, silt and clay, in the soil in the same way as described above. But in determining the soil's textural class he makes use of the triangular graph (Fig. 6.2). Each side of the triangle is graduated from 0 to 100 representing possible levels of sand, silt, or clay in the soil. Thus, if a surveyor estimates that the silt content of his soil sample is 40 per cent and the clay content 30 per cent, the points representing these levels on the scales are located and where the lines projected from these points meet will indicate the soil's textural class; in this case it is clay-loam.

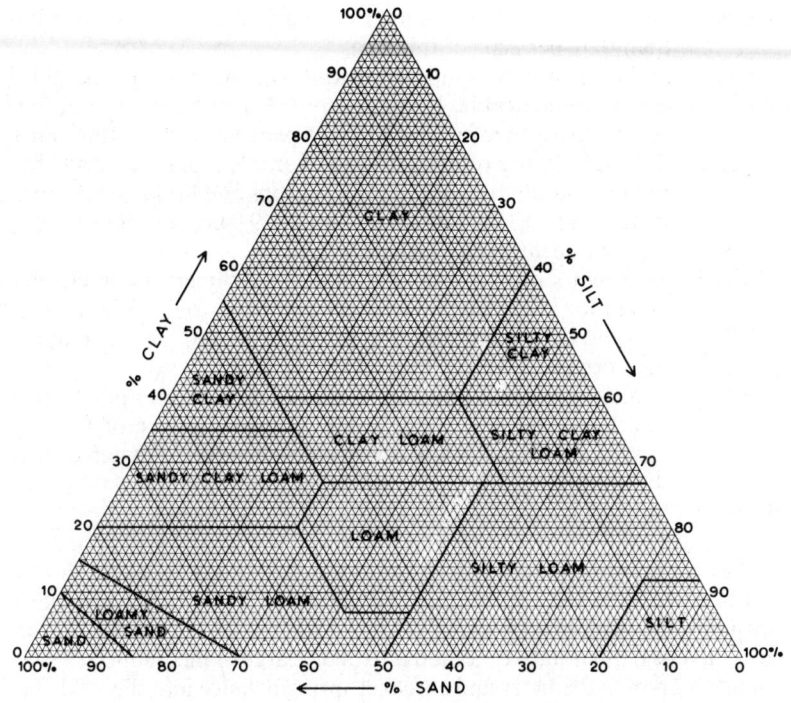

Fig. 6.2 Soil textural classes

COARSE FRAGMENTS (STONES)

These are the unweathered or partly decomposed rock materials in the soil which are larger than 2 mm in diameter. The coarse content of the soil is an important parameter not only because it affects a soil's usefulness for agricultural purposes but also because it is often used to subdivide soil series into phases. With regard to these coarse fragments, four variables are important and should be investigated: (1) abundance and distribution through the soil profile, (2) size, (3) shape, and (4) mineralogical composition and chemical state.

Abundance Coarse fragments up to 254 mm in length are regarded as forming part of the soil mass and they are not supposed to interfere with farming practices. The parameter 'stoniness' usually refers to the relative proportion of stones larger than 254 mm in and on the soil. Soils are grouped into classes of stoniness on the basis of the number of stones, the distances between them, and the proportion of the surface area they occupy.

Class 0 No stones, or too few to interfere with tillage. Stones cover less than 0·01 per cent of the area.
Class 1 Sufficient stones to interfere with tillage but not to make tilled crops impracticable. If stones are 0·3 m in diameter and 9·0–30·5 m apart, they occupy 0·01–0·1 per cent of the surface area.
Class 2 Sufficient stones to make tillage of arable crops impracticable, but the soil can be worked for hay crops and improved pasture. If stones are 0·3 m in diameter and 1·5–9·0 m apart, they occupy 0·1–3 per cent of the surface.
Class 3 Sufficient stones to make all use of machinery impracticable except for very heavy machinery and hand tools. Wild pasture or forests. If stones are 0·3 m in diameter and 0·8–1·5 m apart, they occupy 3–15 per cent of the surface.
Class 4 Sufficient stones to make all use of machinery impracticable. The land may have some value for poor pasture or forestry. Stones are 0·8 m or less apart, and occupy 15–90 per cent of the surface.
Class 5 Stones occupy over 90 per cent of the surface.

For the gravelly material (2–254 mm) it is customary to take a shovelful of earth from the appropriate soil layer and by fingering through it estimate the relative proportion of coarse fragments in the soil. A more objective and quantitative method is to take a large or medium-sized sieve of mesh 2 mm in diameter and a pocket spring balance into the field. The abundance of stones is then measured by:

1 Weighing a sample with the spring balance.
2 Sieving the soil to remove all particles smaller than 2 mm.
3 Weighing the sample again to obtain the weight of coarse material in the soil.
4 Calculating the weight percentage of stones in the soil as

$$\text{wt \% stones} = \frac{\text{wt of coarse material (3)}}{\text{wt of bulk sample (1)}} \times 100.$$

It is sometimes difficult to use this method in the field. For instance the soil may be so wet as to be impossible to sieve. In that case it is advisable to obtain samples in the field which can then be dried and sieved in the laboratory.

Stone size The size of a stone is its length along its major (longest) axis. This can be measured in millimetres or centimetres with the normal graduated rule. The tables on the next page show two schemes for the classification of stone sizes (Tables 6.4, 6.5).

TABLE 6.4
Stone Class Determination

Length	Stone class
3–6 mm	Gravel
6–13 mm	Coarse gravel
13–25 mm	Very small stones
25–50 mm	Small stones
50–100 mm	Medium stones
100–200 mm	Large stones
Over 200 mm	Boulders

After G. R. Clarke (1957).

TABLE 6.5
Names Used for Coarse Fragments in Soils

Shape and kind	Size and name of fragments		
	Up to 7·6 cm in diameter	7·6–25·4 cm in diameter	More than 25·4 cm in diameter
Rounded and sub-rounded fragments (all kinds of rock)	Gravelly	Cobbly	Stony (or bouldery)
Irregularly shaped angular fragments: Chert	Chert	Coarse cherty	Stony
Other than chert	(Angular) gravelly	Angular cobbly	Stony
Thin, flat fragments	Up to 15 cm in length	15–38 cm in length	More than 38 cm in length
Thin, flat sandstones, limestone, and schist	Channery	Flaggy	Stony
Slate	Slaty	Flaggy	Stony
Shale	Shaly	Flaggy	Stony

From USDA (1951).

The USDA scheme recognizes only three broad size classes but distinguishes between different stone materials—chert, sandstone, schist, limestone. The classification by Clarke, though less comprehensive, is simpler and easier to use.

Stone shape The description involves not only a consideration of the geometric structure of the stone but also the ruggedness of the edges of the stone. The following terms are commonly used:

Angular	Sharp pointed edges
Subangular	Slightly less sharp edges
Blocky	Tabular or rectangular in shape (variants—angular and subangular blocky)
Rounded	Smooth edges, like nodules
Platy or shaly	Flat (variants—angular and subangular platy)

Mineralogy and chemical state The various types of rock fragments—granite, quartzite, shale, etc.—present in the soil should be recorded as well as their mineralogical composition. The chemical state of the rocks is also investigated. For example, some of the stones may be partly decomposed and crumble easily; some are hard and resistant; still others are secondary concretionary deposits (like iron concretions) which have accumulated in the soil.

STRUCTURE AND CONSTITUTION

Structure refers to the characteristic arrangement of the soil particles. There are three ways of describing soil structure:

Type or shape of peds (i.e. soil aggregates)
Size of peds
Strength, i.e. the tenacity or resistance to pressure of the peds

Type and size The following terms are used in connection with type and size of peds:

Crumbly	This is a term used in describing the structure of agricultural soil horizons which have been modified by farming practices. Soil crumbs are small, blocky, or nutty soil peds, usually 2–5 mm in diameter, which are also very porous. Sometimes a distinction is made between 'water stable'

and 'water unstable' crumbs. The latter crumble easily under gentle drops of falling water while the former do not. Soil crumbs join together to form clods. These are irregularly shaped aggregates which are less porous than the crumbs.

Granuole or granular Hard, almost non-porous spheroidal, 1–4 mm in diameter.

Prismatic Prism-like peds in which the vertical axis is much longer than the horizontal axis. The peds are bounded by flat surfaces and they may be over 10 cm in diameter.

Columnar The peds are also like prisms but they have rounded caps; they also may be over 10 cm in diameter.

Blocky Block-like peds bounded by flat or rounded surfaces. There are two subtypes: (1) angular blocky—peds in which the flat surfaces intersect at very sharp angles; they are fairly porous but are longer than crumbs (over 50 mm in diameter); (2) subangular blocky—peds which have both flat and rounded surfaces (may be over 50 mm in diameter).

Platy Flat peds in which the horizontal axis is much longer than the vertical. Some of these peds are very fine indeed, being less than 1 mm thick, but there are some which may be over 10 mm in thickness.

CONSISTENCE

The degree of cohesion of the soil mass and its resistance to pressure is known as the soil consistence. Consistence varies with the moisture condition of the soil, whether dry, moist, or wet.

Consistence when dry Soils are most resistant to pressure when dry. The soil is characterized by rigidity or brittleness and a tendency to break down to a powder or fragments with sharp edges. The categories are:

0	Loose	non-coherent soil mass
1	Soft	weakly coherent and fragile, breaks to a powder or individual grains under very slight pressure
2	Slightly hard	weakly resistant to pressure; easily broken under slight pressure
3	Hard	moderately resistant to pressure; barely breakable between the thumb and the forefinger
4	Very hard	very resistant to pressure, not breakable between fingers and can be broken in the hand only with difficulty

Consistence when moist The soil mass is neither rigid nor brittle; it tends to break down into smaller masses rather than into powder and the material can be pressed together again after disturbance. The categories are:

0	Loose	non-coherent soil mass
1	Very friable	soil mass crushes easily under very gentle pressure but coheres when pressed together
2	Friable	soil mass crushes easily under gentle to moderate pressure and coheres when pressed together
3	Firm	soil mass crushes under moderate pressure but resistance is distinctly noticeable
4	Very firm	soil mass crushes under strong pressure, barely breakable between the thumb and the forefinger
5	Extremely firm	soil mass crushes only under strong pressure; cannot be crushed between the thumb and the forefinger

Consistence when wet Wet soils tend to be sticky and plastic, i.e. they have the quality of adhesion to other objects and the ability to change shape continuously under the influence of an applied stress and to retain the impressed shape on removal of the stress. This assumes, of course, that the wet soils are rather clayey.

POROSITY

This is determined by the size and number of pore spaces, cracks, and fissures in the soil. Porosity is described in the field under two headings: (1) aggregate porosity, i.e. the size, shape, and distribution of spaces within soil aggregates; and (2) the nature (size, shape, and distribution) of spaces between aggregates.

Aggregate porosity is defined in terms of a size name, shape and distribution as follows (Clarke, 1957):

Size name		Shape and distribution	
Fine porous	1 mm diameter	Honeycomb	Horizontal
Porous	1–3 mm diameter	Vermiculate	Vertical
Spongy	3–5 mm diameter	Rounded	Oblique
Cavernous	5–10 mm diameter	Polygonal	Ubiquitous
		Dendritic	

Between aggregates, or the relationship between the shape of spaces between soil aggregates and the shape of the aggregates themselves, are described as:

Very fine-fissured (spider web)		Horizontal
Fine-fissured	1 mm wide	Vertical
Fissured	1–3 mm wide	Oblique
Wide-fissured	3–5 mm wide	Ubiquitous
Very wide-fissured	5–10 mm wide	

SOIL MOISTURE AND DRAINAGE CHARACTERISTICS

A soil can be described qualitatively as dry, slightly moist, very moist, damp, or wet, depending on the amount of water in it. Generally, soil moisture determination is of two types. First, there are those methods designed to measure the total soil moisture capacity, e.g. the tensiometer (or suction) method. Second, there are those techniques, such as the oven-drying method, the gypsum block, and the neutron scattering methods, which measure the actual amount of water in the soil. To obtain an accurate picture of soil moisture characteristics at all times with these techniques, moisture content measurements have to be made over a period of time or at different seasons of the year, as soil moisture content varies from time to time within a day and between seasons. For measuring soil moisture during the course of the field trip the gypsum block method is very handy and useful.

The equipment consists of two small blocks of gypsum tied to the ends of two strings (negative and positive) of electric conductors which are connected to a meter which records electrical resistance. The gypsum blocks are dipped into the soil and can be left for as long as is required. The blocks absorb moisture from the soil and their resistance to electrical current is related to the amount of moisture absorbed. The electrical resistance is read on the recording meter. The moisture content is then read off a standard curve (graph) prepared beforehand by calibrating resistance readings with soil moisture content.

Soil drainage depends on its moisture-to-air ratio which in turn depends on many factors, as we have discussed in Chapter 2. Although soil drainage can be classified into many grades depending on the moisture-to-air ratio and the availability of moisture for plant growth, these are difficult to assess in the field. The major distinction which is easily observable in the field is between well-drained soils on the one hand and poor to imperfectly drained soils on the other. In well-drained soils water moves easily and the iron compounds are fully oxidized so that there are no pale or reduction colours (i.e. no gleying) or mottling. The soil horizons also tend to be more uniformly coloured. Poorly drained soils have an excessive moisture-to-air ratio; anaerobic conditions produce reduction colours and possibly mottling. In these poorly drained soils the chief interests lie in the depth below the surface at which gleying commences, and in the pattern of mottling. The depth at which gleying occurs can be measured with any linear measuring instrument.

With regard to mottling two important aspects studied in the field are (1) the colour of the soil matrix and of the principal mottles, and (2) the pattern of the mottling. Colour can be determined using the Munsell colour chart as explained earlier in this chapter, while three variables—contrast, abundance, and size—constitute the pattern of mottling.

Contrast is the distinctness of the mottle colour(s) from that of the soil matrix. It can be described as:

Faint   Indistinct mottles which can be recognized only after close examination. The colours of the mottles and the soil matrix have closely related hues and chroma.
Distinct   The mottles have hue, value, and chroma readily distinguishable from those of the soil matrix. They are therefore readily observable.
Prominent   The mottles are conspicuous and mottling is one of the outstanding features of the horizon. The hue, value, and chroma differ markedly from those of the soil matrix.

Abundance is defined as the proportion of the exposed area of the horizon that is occupied by mottles. It can be described as:

Few   Mottles occupy less than 2 per cent of the exposed surface.
Common   Mottles occupy 2–20 per cent of the exposed surface.
Many   Mottles occupy more than 20 per cent of the exposed surface.

Size is expressed by the approximate diameters of individual mottles, as:

Fine   Mottles less than 5 mm in diameter (along the greatest dimension).
Medium   Mottles range between 5 and 15 mm in diameter.
Coarse   Mottles are greater than 15 mm in diameter.

ORGANIC MATTER

Under this heading are included the description of the litter or plant residue on the soil surface and the nature and distribution of dead and living plant roots, as well as the nature and disposition of the organic matter in the soil profile.

Litter   Four aspects are to be investigated and measured if possible:

The thickness of the litter layer which can be measured using a graduated rule.

The layering of the litter.
The composition of the layer, including leaves, twigs, tree trunks, fruits, animal remains, etc.
The stage of decomposition reached by the litter.

Plant roots The pattern of distribution of roots in the soil is often an index of some important soil characteristics such as texture, structure and consistence, and moisture condition. Clarke suggests seven different ways of describing and classifying plant roots out of which we have chosen three which we consider to be important. These are:

a. Quantity, estimated as:
Abundant—more than 100 roots per 1,000 sq cm of profile/horizon face
Frequent—100–20 roots per 1,000 sq cm of profile/horizon face
Few—20–4 roots per 1,000 sq cm of profile/horizon face
Rare—3–1 roots per 1,000 sq cm of profile/horizon face
b. Size, described as:
Large—roots more than 12 mm in diameter
Medium—12–3 mm in diameter
Small—3–1 mm in diameter
Fine—1 mm in diameter
c. Nature, described as woody, fleshy, fibrous, or rhizomatous.

Humus Four main types of humus may be recognized in the field, namely,

a. Raw humus (mor) or thick accumulation of partly decomposed organic matter on the soil surface. The raw matter is soft when waterlogged but it hardens, shrinks, and cracks when dry. The layer of raw humus is usually sharply demarcated from the mineral horizon below it.
b. Mild humus (mull) or well decomposed, finely divided organic matter which is intimately mixed with the soil mineral particles.
c. Intimate humus or humus particles or layers found not only in the A horizon but also dispersed in the B horizon. It is the humus truly incorporated with the soil mass producing black, brown, and grey colours. They exist as discrete particles, as coating on soil minerals, and as cement between mineral particles.

It is possible in the field to actually determine experimentally the amount and depth of humus distribution in the profile by using hydrogen peroxide. A little hydrogen peroxide is poured down a freshly cut face and the degree of effervescence is noted. Where effervescence stops down the soil profile is the limit of humus penetration.
d. Mechanically incorporated organic matter, including dead root and

plant material washed down mechanically into the soil through cracks and animal and root channels. These are later decomposed and humified to become intimate humus.

Faunal influence Following closely on the description of soil organic matter should be the consideration of the influence of soil fauna such as earthworms, eelworms, termites, ants, and moles, etc. Their visible effects on the soil include burrowing and the presence of anthills, molehills, and wormcasts. Earthworms are a good indicator of soil reaction and aeration because they are very sensitive to acidity and anaerobic soil conditions.

SOIL pH

This can be measured potentiometrically or colorimetrically in the field using either a portable battery-operated pH meter or indicator dyes. The pH measurements can be distorted if sufficient care is not taken to prevent contamination of any kind from perspiration, dust particles, and chemicals, among other things. It is advised that the soil sample intended for use should not be touched by hand at all (as sweat may cause an acidic reaction) and should be put in a very clean container. Distilled water should be used throughout for making the soil solution and for cleaning the equipment after use.

Potentiometric method

1. Put a small sample of soil (approx. 5 g) in a receptacle.
2. Add enough distilled water (about 25 ml) to saturate the sample and stir the solution very well. Allow to stand for some time.
3. Stir the solution again and then immerse the two electrodes of the pH meter in the solution. The pH is read on the recording meter.

In order to obtain accurate results the pH meter has to be checked and set or 'buffered' so that it is in good working order. When the meter is switched on the two electrodes are immersed in distilled water. Then, using the 'set zero' knob, the recording dial is set to zero on the pH scale. The two electrodes are then immersed in a 'buffer solution' of known pH, say 4·0. The 'adjust' knob is used to set the dial to 4·0 on the meter scale. The electrodes are sprayed with distilled water and then immersed in a second buffer solution with a pH of, say, 7·0. If the meter is in good order the dial should read 7·0 on the meter scale. If it does not, the 'adjust' knob is used, and the process repeated until the meter is well set. All pH meters have a temperature scale for use in cases where the electrodes have no built-in temperature control. Before the pH of the solution is measured its temperature is taken and the temperature of the pH meter is adjusted accordingly.

Colorimetric method The method is much cheaper, quicker, and easier to use than the potentiometric method. A variety of indicators are used which give near accurate results each within certain pH ranges. The one that is used in any given area will therefore depend on the range of pH values expected. The indicators themselves can be chemically unstable and may distort results. However, it is believed that they give satisfactory results for soils with pH of between 4·5 and 7·5.

A commonly used dye is the BDH Universal Soil Testing Indicator. There is a colour chart and a small elongated container to go with the kit. The procedure for determining pH is as follows:

1 Put a small sample of soil in the wider end of the container.
2 Add enough indicator to saturate and immerse the sample completely. Allow to stand for about a minute for the reaction to take place. Barium sulphate powder with pH 7·0 can be added to bring out the colour of the liquid which may otherwise be masked by the soil colour.[1]
3 Tilt the container slightly to separate the solution from the sample. Compare the colour of the solution with the colours of the colour chart to find the closest match.

FREE CARBONATE

The concentration of free carbonates in the soil is tested by pouring a small quantity of acid on the soil sample and noting how audible and how strong the effervescence is. Table 6.6 gives the possible measures.

SECONDARY MINERALS

Many of these can only be identified in the laboratory. The observable minerals in the field include concretionary deposits, for example, of iron, aluminium, and manganese; and fine salt precipitates like calcium carbonate and gypsum.

Soil Monolith

The collection of small soil samples in polythene bags and envelopes from different horizons has been mentioned earlier in this chapter. But on

---

[1] There are many ways of preparing the sample for pH test with indicator dyes. One method is to take about a gramme of soil, put it on a porcelain plate, and add enough indicator to make a thick paste. After mixing thoroughly barium sulphate is sprinkled on. The colour is read after some time.

Another method involves first adding distilled water to the soil samples with barium sulphate in a test tube. The solution is stirred and allowed to stand for about five minutes before adding a few drops of indicator.

TABLE 6.6
Determination of Percentage Free Carbonate Content

% Free carbonate	Audible effects	Visible effects
0·1	None	None
0·5	Faintly audible increasing to slightly	None
1·0	Faintly audible increasing to moderate	Slight effervescence confined to individual grains; just visible
2·0	Moderate to distinct, heard away from ear	Slightly more general effervescence visible at close inspection
5·0	Easily audible	Moderate effervescence, bubbles to 3 mm, easily visible
10·0	Easily audible	General strong effervescence, ubiquitous bubbles to 7 mm, easily visible

From G. R. Clarke (1957).

occasions, and for record and teaching purposes, one may want to take and preserve in the laboratory a whole soil profile as found in the field. This sample soil profile is known as a 'soil monolith'. The collection of the monolith requires much labour and care but these are well rewarded by the opportunity it affords of studying the soil profile in its natural state in the comparatively comfortable surroundings of a laboratory or classroom.

The monolith can be made of wood or steel, but the latter is stronger, more durable, and easier to insert into the soil in the field than the former. Monoliths vary in size depending on the thickness of the soil profiles and on whether a continuous or a discontinuous representation of the soil profile is wanted. In the discontinuous monolith a small representative section of each of the soil horizons is cut and the pieces are then fitted together to form a monolith. This method considerably reduces the length of the monolith.

The complete monolith unit consists of (1) an open-ended trough, (2) a lid, and (3) base cover (see Taylor, 1960). Also required are a sharp knife, a spade, and vinyl resin.

To Extract the Monolith

1. Prepare the soil profile pit and choose the side from which the monolith is to be extracted.
2. Mark out the area of the profile face required.
3. Using a sharp cutting instrument (spade or a steel blade), cut the two vertical sides and the bottom end of the block.
4. Place the open-ended trough against the profile face and push or knock it into the soil through the cuts.[1] To help the insertion of the trough, the profile walls on either side can be cut back.
5. When the trough has been pushed well in, the back end of the block of sample is cut and the monolith base cover is inserted and nailed or screwed on. The lid is also inserted and screwed.
6. To protect the surface of the monolith from drying out and losing its natural colour and structure, the soil monolith is impregnated with vinyl resin.

---

[1] In soils with an indurated layer, it may be practically impossible to knock the open-ended trough into the soil. In such a case, the cutting of representative blocks from each soil layer and fitting them together to make a discontinuous monolith would be more feasible.

Chapter 7

SOIL CLASSIFICATION

To many people this is where soil geography begins. To them geographers are essentially interested in the distribution of soils over the earth's surface. Yet it is not difficult to see the weakness of this viewpoint. Soil description and classification will be meaningless, if at all possible, without a proper knowledge of the soil, its properties, and constituents, the environments in which it has developed; the processes involved in its formation; and the standard techniques used in its study. In other words, the earlier chapters in this book are necessary preliminaries for the classification and intelligent description of soils.

Like all natural phenomena, soil distribution on the globe conforms to some basic laws or patterns. These laws are commonly described as the principles or laws of natural distribution. They include:

(a) The principle of graded likeness and infinite differences. This means, with respect to soil, that no two soil profiles are exactly alike, although similarities may exist which may permit grouping or classification of some sort. So, although minute differences can be observed that are difficult to integrate into a rational, systematic study, yet there are regularities or order within the 'chaos' which permit of some form of classification.

(b) The principle of areal transition. This states that the change in the characteristics of natural phenomena from one place or area to another is always gradational, although the rate of change may vary. The transition becomes more gradual with closeness of scrutiny. This means that the lines drawn in a classification exercise are mainly functional, empirical, and unreal. Such lines are generalizations of areal transitions and more often than not are arbitrary. This idea has been mentioned in an earlier chapter on soil formation processes. Although certain processes are recognized in certain areas, the boundary between two adjacent processes cannot be precisely located, except by definition. The point at which podzolization stops to give way to calcification, or calcification to salinization, etc., cannot be located precisely. The solution to this problem of boundary location in climatology, for example, is essentially empirical, as evidenced by the work of Koppen, Thornthwaite, and others.

(c) The principle of continuous alteration of forms with time. This principle implies that because energy is always moving (flowing) and acting on the forms (e.g. soils) of the earth's surface, and, perhaps more importantly, because the flow of energy in question is mostly irregular, both long- and short-term changes are taking place among the forms and phenomena. With respect to soil, this principle says that the process of soil formation is continual, if not continuous. The soil is always changing. Erosion at the surface and addition to the profile at depth are examples of the changes taking place all the time. This means that to observe the short-term changes we need detailed studies at the micro-level, while long-term changes can be observed in the larger forms and phenomena.

(d) Finally, there is the principle of development towards some sort of balance, equilibrium, or steady state in a given environment. That is, although change continually takes place within the many elements that make up a natural body, periods of relative stability and marked mutual adjustments soon set in to produce a tendency towards a steady state condition. This is identical to the concept of the mature or equilibrium soil, in which soil formation and soil destruction occur simultaneously, but at such a balanced and slow rate as to cause no perceptible change in the soil system (Nikiforoff, 1942, 1943).

Soil classification is therefore necessarily a simplification, an attempt to introduce some order into the chaotic situation of complex and infinite variety of differences both in time and in space. Such an order or simplification is necessary for understanding; it also permits detailed study of the parts as well as facilitating comparison both between parts and between these and the whole. In other words, classification provides a basis for further inquiry. As more information becomes available, the system undergoes modification, as shown later in this chapter.

The Basis of Soil Classification

Soil, as pointed out earlier, is being studied by many branches of science, and each branch has its own approach. There are therefore many systems of soil classification. The most common practice is to arrange soils into assemblages according to certain selected properties. The properties selected are a function of the interest of the classifier, and different classifications emerge from the use of different classificatory criteria. There are systems which are more relevant to geographers and others to agronomists, engineers, etc.

The schemes currently in use can be discussed under four main headings: (a) the empirical or taxonomic approach, using properties such as texture, base content, etc.; (b) the morphologic approach based on profile characteristics as observed in the field; (c) the genetic approach, using certain formative or environmental factors; and (d) the integrative approach, involving varying combinations of a, b, and c above. The empirical and morphological systems are also often combined as the generic approach, but they are distinguished here because of the different areas of emphasis. We shall discuss each of these approaches briefly.

Empirical Classification of Soil
This is the oldest system of soil classification, in which the properties that appeared to be significant in the use of virgin soils for crop growth were used. This was long before the advent of organized soil study, or even that of modern agronomic science. Since this system featured prominently in the US studies, we may use that country as an example here.

Two major properties of the soil featured prominently in the US soil surveys. These were the texture of the topsoil and the type of parent material. The texture of the topsoil was used primarily because the general ease of cultivating a soil was closely related to it. It was also perhaps the most important as far as soil fertility and productivity, including response to treatment, were concerned. The parent material, for its part, was considered to be the determinant of soil texture, and so of the fertility of virgin soils.

This early system of soil classification is still being used, particularly at the local level, e.g. at the series and type levels, especially the latter. Soil series are distinguished on account of profile characteristics, and so belong to a later section. Soil types, on the contrary, are distinguished by the surface textural class, e.g. silt, silt-loam, sandy loam, loam (see Chapter 6).

Soils may also be classified by any other of the observable properties, e.g. colour, depth, pH, base exchange capacity, nutrient content, clay content, etc. For instance, the soils of a part of the Sydney district of New South Wales (Australia) have recently been classified by colour into: (a) brown podzolic soils developed on predominantly indurated zone material; (b) yellow to red podzolic soils developed on predominantly mottled-zone material; and (c) grey and white podzolic soils developed on pallid zone and other non-ferruginized, non-oxidized material (Faniran, 1968). These soil types were observed to occur in that order from many hilltops, plateau-tops and other interfluves to the adjacent river valleys of the erosionally dissected countries. They indicate pedogenesis consequent on the dissection of a deeply weathered and duricrusted surface. Similar catenary sequences have been described in many parts of Australia, as described in Chapter 4. Observations in Nigeria suggest similar situations but these have not been studied specifically.

Finally, classifications based on the chemical content of soils—e.g. pH, base content, etc.—abound in the agronomic literature. These factors directly influence soil fertility and are given prominence in soil productivity, soil capability, and soil potentiality investigations (see opposite).

Morphological Classification of Soil

The main aim of this system is to infer genesis from profile characteristics. As such it is basically subjective, since profile morphology does not always reflect the processes of soil formation. Moreover, this system is frequently mixed up with the empirical system in that properties such as colour, texture, etc. are used. This is not unexpected, judging from the place in the hierarchy of soil classes it belongs to. As was hinted earlier, morphologic classification of soils is mainly at the *series* level, in which case it groups together a number of similar soil types. The American (Seventh Approximation) system, which is basically a morphological classification of soils, however, recognizes similar profile characteristics even at the *order* level, as shown later in this chapter.

TABLE 7.1
World Soils According to G. W. Robinson (1947)

Freely drained soils	Pedalfers (completely leached)	Presence of raw humus	1 2	Humus podzols Iron podzols
		Absence of raw humus	3 4 5 6 7 8 9 10	Brown earth Degraded chernozems Prairie soils Yellow podzolic soils Red podzolic soils Tropical red soils Ferrallites Chernozems
	Pedocals (incompletely leached)		11 12 13	Chestnut soils Brown desert soils Grey desert soils
Soils with impeded drainage	Absence of soluble salts	Subarctic	14	Tundra
		Temperate	15 16 17 18	Gley soils Gley podzols Peat podzols Peat soils
	Presence of soluble salts	Subtropical and tropical	19 20 21 22	Vlei soils Saline soils Alkaline soils Soloti soils

Modified from Bridges, 1970, p. 27.

Another system of soil classification based at least partly on profile or morphological characteristics of soils is that proposed by Robinson (1949). According to this system, world soils are classified according to the nature of leaching and the material found in the B horizon. Two major categories are recognized, namely, pedalfers and pedocals. Pedocals contain a lime accumulation horizon while pedalfers contain a horizon enriched by the oxides of iron and aluminium. Each of the two categories is characteristically subdivided in a hierarchical system (Table 7.1).

Genetic Classification of Soil
This is the system introduced by the Russians late in the nineteenth century. The system is based largely on soil characteristics, especially those which reflect the active processes of soil formation, i.e. climatic and biological soil processes. The system is different from the previous ones in being genetic rather than generic. We have discussed the main workers in this connection, particularly Dokuchaev and Sibirtzev, the pioneers of pedology (Chapter 1). Also in connection with the genetic approach to soil study and classification, the activities of Marbut are outstanding. By successfully introducing the Russian system to America, and also improving on the system, he was instrumental in its spread throughout the English-speaking world at least (cf. Marbut, 1935).

The classification system introduced by Marbut into the USA from the original Russian (Sibirtzev) version comprised three major orders—zonal, intrazonal, and azonal soils. Zonal soils included those whose characteristics are mainly the result of climatic or biological conditions. Intrazonal soils were considered to be those soils whose properties reflect local conditions of drainage or parent material more than they reflect the climate or vegetation. Azonal soils included miscellaneous unconsolidated materials that have not yet developed distinct soil horizons (Table 7.2). These major categories are then subdivided into suborders and great soil groups, etc., in a way similar to botanical or zoological classification (cf. Table 7.3).

Now let us try to define some of the terms being used in genetic soil classification. The most important ones are order, suborder, great soil groups, association, series, type and phase.

Soil orders usually relate to the five soil-forming factors discussed in Chapter 4. But as used in many systems of soil classification, the soil order is the highest unit in soil classification and usually refers to the tripartite division into zonal, intrazonal, and zonal soils.

Soil suborders are generally associated with the very obvious features of soil profiles, e.g. domination by organic matter, accumulation of lime or sesquioxides, marked illuviation of constituents, or the accumulation of salts. Both zonal and intrazonal soils have suborder representatives but azonal soils have not been so distinguished.

TABLE 7.2
Classification of Soils: The Great Soil Group System

Order	Suborder	Great soil group
Zonal soils	Soils of the cold zone	Tundra Podzol
	Light-coloured podzolized soils of timbered regions	Brown podzolic soils Grey-brown podzolic soils Red-yellow podzolic soils Grey podzolic or grey wooded soils
	Soils of forested warm-temperate and tropical regions	A variety of latosols are recognized; they await detailed classification
	Soils of the forest-grassland transition	Degraded chernozem soils Non-calcic brown or shantung brown soils
	Dark-coloured soils of semi-arid subhumid and humid grassland	Prairie soils (semipodzolic) Reddish prairie soils Chernozem soils Chestnut soils Reddish chestnut soils
	Light-coloured soils of arid regions	Brown soils Reddish-brown soils Sierozem soils Red desert soils
Intrazonal soils	Hydromorphic soils of marshes, swamps, flats, and seepage areas	Humic-glei soils (includes wiesenboden) Alpine meadow soils Bog soils Half-bog soils Low-humic glei soils Planosols Groundwater podzols Groundwater latosols
	Halomorphic (saline and alkali) soils of imperfectly drained arid regions, littoral deposits	Solonchak soils (saline) Solonetz soils (alkali) Soloth (solodi) soils
		Brown forest soils (braunerde) Rendzina soils
Azonal soils	No suborders	Lithosols Regosols (includes dry sands) Alluvium

From Thorp and Smith, 1949.

TABLE 7.3
A Comparison of Botanical and Soil Classification Systems

Botanical		Soil	
Class	Example	Class	Example
Phylum	Spermatophyta	Order	Zonal
Order	Angiospermae	Suborder	Podzol
Class	Dicotyledoneae	Great soil group	Red podzolic
Family	Fagaceae	Family (America)	Miami
		Association (Australia)	Cumberland
		Fasc (Nigeria)	Iwo
Genus	Overcus (America)	Series	Hillsdale (America)
	Eucalyptus (Australia)		Iwo (Nigeria)
			Cumberland (Australia)
			Miami (America)
Species	Obliqua (Australia)	Class (America)	Miami loam
	Alba (America)	Type (Nigeria)	Mamu loam
	Acacia spp. (Nigeria)		
Common name	Messmate (Australia)		
	White oak (America)		

The great soil groups are based on total profile features. As was pointed out earlier, these features or properties reflect the formative factors and processes, particularly the latter. These processes and associated properties form a continuum. The great soil groups therefore frequently grade one into another. The named examples are only nodes along the long continuum of variations in soil properties. It is important to note also that the names used in the great soil groups are pedological language, which are difficult to eliminate from the literature. The great soil groups are sometimes subdivided into great groups and often into soil associations or soil families as the case may be. These soils differ in the structure, colour, or the degree of mottling of the B horizon. In the podzol group, for example, we may recognize soils with a thick, soft Bs horizon or those with a thin, hard Bs horizon. This is the pan or hardpan layer of sesquioxide accumulation. Similarly, in the chernozem group, we may distinguish between 'leached', compact, or typical chernozems.

The term soil association was first used by Ellis in 1932. It was then defined as a group of 'topographically related soils developed on one geological parent material'. As such it refers to what is generally referred to as a topo-sequence of soil series or a catena sequence (see Chapter 4). It is generally given a geographic name, e.g. after a typical area of occurrence (Table 7.3).

The soil series is a subgroup of the soil association. It is a group of soils having horizons similar in the differentiating characteristics and arrangement in the profile. This does not normally include the texture of the surface soil but does refer to soils formed in the same parent material. Soil series, like soil associations, have the same range of colour, texture and, horizon sequence, and the same conditions of relief and drainage. They also have similar site, origin, and mode of formation. Also, like a soil association, a soil series is usually given a geographic name from the area in which it was first described, e.g. the Iwo series. In addition, it has a textural designation. Both soil association and soil series constitute the most frequently used soil-mapping units.

The next lower-order category of soils is soil type. This is conventionally explained in terms of textural variations of the A horizon within a soil series. Thus several textural variants are assembled to form a series. Slope is an important factor here. Finally, the soil type may be distinguished into soil phases, which reflect erosional state and so express variations of depth, stoniness or disruption of the A horizon. Soil phases are used mainly in Britain.

Integrative Systems of Soil Classification

Opinion is still widely divided among pedologists as to which of the above-described systems is best to use. Consequently, most of the systems of soil classification in use today attempt to combine features of the three approaches. Soils are commonly arranged in a hierarchical order or an upward and integrative assemblage, say from soil type or soil series to soil order (cf. Tables 7.2 and 7.3).

In order to illustrate this approach, a few of the available systems of soil classification will be described. The fact that we have many systems shows that soil classification is not as perfect as either botanical or zoological classification. The search for a standard, satisfactory system still continues. The following illustrate the attempts already made in this direction.

THE AUSTRALIAN SYSTEMS OF SOIL CLASSIFICATION

At least four systems of soil classification have been proposed by Australian soil scientists. These include those by Stephens (1962), Northcote (1965), Stace and others (1968), and by Corbett (1969) (cf. Tables 7.4 and 7.5). The similarities between Stephens's system (Table 7.4) and Table 7.2 are apparent. According to Stephens's usage, soil orders depend on whether lime and/or gypsum are found in the solum (cf. Robinson's (1949) system (Table 7.1)). Also, like the Robinson system, his suborders depend on the most obvious profile features, e.g. organic matter, clay, sesquioxides, etc.

The systems proposed by Northcote (1965) and by Stace and others (1968) differ from that of Stephens and also from each other. Northcote's classification sets aside the 'great soil group' concept, aiming at 'precise definitions of the soil profile'. Using a coding system, he arrives at 500 soils for Australia. Four primary profiles are recognized as:

O soils dominated by organic matter
U uniform primary profile forms: soils dominated by mineral matter but showing little textural change down the profile, i.e. no clearly defined textural boundaries within the solum
G gradational primary profile forms: soils dominated by mineral matter and becoming increasingly finer in texture with depth
D duplex primary profile forms: soils dominated by mineral matter, showing marked textural contrast between the eluvial and illuvial horizons.

In addition, Northcote recognizes a number of field texture classes:

Uc coarse textured
Um medium textured
Uf fine textured, non-cracking
Ug fine textured, cracking

This is definitely a very complex system.

TABLE 7.4
Soil Classification According to Stephens (1962)

I SOLUM CLASS—UNDIFFERENTIATED
1 Alluvial soils
2 Skeletal soils
3 Calcareous coastal sands

TABLE 7.4 (contd)

II SOLUM CLASS—DIFFERENTIATED	

Pedalfers
a Solum dominated by acid peat or peaty eluvial horizon
    4 Moor peats
    5 Alpine humus
    6 Moor podzol peats
    7 Acid swamp soils
b Solum acid, and with organic sesquioxide and sometimes clay illuvial horizons
    8 Podzols
    9 Groundwater podzols
c Solum acid and with clay and sesquioxide illuvial horizons
    10 Lateritic podzolic soils
    11 Grey-brown podzolic soils
    12 Brown podzolic soils
    13 Red podzolic soils
    14 Yellow podzolic soils
    15 Meadow podzolic soils
    16 Non-calcic brown soils
d Solum acid to neutral and lacking pronounced eluviation of clay
    17 Yellow earths
    18 Krasnozems
    19 Lateritic krasnozems
    20 Lateritic red earths
    21 Terra rossa
    22 Prairie soils
 	
Pedocals
e Solum dark-coloured and slightly acid to neutral in eluvial horizons; calcareous illuvial horizons
    23 Black earths
    24 Wiesenboden
    25 Brown forest soils
    26 Rendzinas
    27 Groundwater rendzinas
    28 Fen soils
	f Solum saline or showing post-saline structure in the illuvial horizon
	    29 Solonchaks
	    30 Solonetz
	    31 Solodized-solonetz
	    32 Soloths
	    33 Solonized brown soils
	g Solum with slightly acid to neutral eluvial horizons; calcareous illuvial horizons
	    34 Red-brown earths
	    35 Brown earths
	    36 Brown soils of light texture
	    37 Arid red earths
	    38 Grey calcareous soils
	h Solum with neutral to alkaline weakly developed eluvial horizons; calcareous and/or gypseous illuvial horizons
	    39 Grey soils of heavy texture
	    40 Brown soils of heavy texture
	i Solum with deflated, slightly acid to alkaline eluvial horizons; calcareous and/or gypseous illuvial horizons
	    41 Desert loams
	    42 Grey-brown and red calcareous desert soils
	    43 Red and brown hardpan soils
	    44 Desert sand plain soils
	    45 Calcareous lateritic soils
	    46 Stony desert tableland soils
	    47 Desert sandhills

	Soil orders are *Pedalfers* and *Pedocals* Soil suborders are *a* to *i* Great soil groups are 1 to 47

TABLE 7.5
The Soil Classification System of Stace, Hubble, and Others (1968)

1 No profile differentiation	1 Solonchaks 2 Alluvial soils 3 Lithosols 4 Calcareous sands 5 Siliceous sands 6 Earthy sands
2 Minimal profile development	7 Grey, brown, and red calcareous soils 8 Desert loams 9 Red and brown hardpan soils 10 Grey, brown, and red clays
3 Dark soils	11 Black earths 12 Rendzinas 13 Chernozems 14 Prairie soils 15 Wiesenboden
4 Mildly leached soils	16 Solonetz 17 Solodized-solonetz and solodic soils 18 Soloths 19 Solonized brown soils 20 Red-brown earths 21 Non-calcic brown soils 22 Chocolate soils 23 Brown earths
5 Soils with predominantly sesquioxidic clay minerals	24 Calcareous red earths 25 Red earths 26 Yellow earths 27 Terra rossa soils 28 Euchrozems 29 Xanthothems 30 Krasnozems
6 Mildly to strongly acid and highly differentiated	31 Grey-brown podzolic soils 32 Red podzolic soils 33 Yellow podzolic soils 34 Brown podzolic soils 35 Lateritic podzolic soils 36 Gleyed podzolic soils 37 Podzols 38 Humus podzols 39 Peaty podzols
7 Dominated by organic matter	40 Alpine humus soils 41 Humic gleys 42 Neutral to alkaline peats 43 Acid peats

SOIL CLASSIFICATION

The system by Stace and others (1968) discards the solum classes and soil orders. Solum classes are combined under their suborder and soil order is not used at all. The system is aimed principally at reflecting degree of leaching (Table 7.5). Finally, Corbett (1969) proposes a much simpler system. For Australian soils she recognizes:

1. Soils dominated by sesquioxides, e.g. laterites and lateritic soils, krasnozem, red earth, yellow earth.
2. Podzolized soils: podzol, groundwater podzol, brown podzolic, brown forest soil, red podzolics, yellow podzolic.
3. Soil dominated by organic matter—alpine humus soil, acid moor peat, humic gley soil, fen soil.
4. Soils dominated by silicate clay—shallow red soil, chocolate soil, prairie soil, black earth, chernozem, terra rossa, rendzina, etc.
5. Soils dominated by base accumulations—red-brown soils, non-calcic brown earth, solonized brown soil, solonchak, solonetz, solodized-solonetz, solodic, desert loam, desert sand, sierozem.
6. Relic soils—lateritic krasnozem, lateritic red earth, calcareous lateritic soil, lateritic podzolic, stony desert soil, red and brown hardpan soil.

Corbett's system appears to be a great improvement on the earlier systems. The equivalences in the other systems are shown in Table 7.6a. The influence of the diagnostic horizon characteristics is here quite obvious.

TABLE 7.6a
The New American Soil Classification System: Summary of the Global Soil Orders

Soil order	Diagnostic properties	Equivalent soils
Entisols	No diagnostic horizons or profile development	Azonal soils, tundra, lithosols, regolith, alluvium, sands
Vertisols	Dark topsoils that crack badly when dry; slanted columnar structure	Grumusols, tropical black clays, regur, tirs
Inceptisols	Young soils, with profile features just beginning; no clear-cut diagnostic horizons, usually dark epipedons	Subarctic brown forest, brown forest, ando, lithosols, regosols, some humic gleys
Aridisols	Desert soils that have one or more diagnostic horizons but no spodic or oxic horizons; light-coloured epipedons	Desert, red desert, grey desert (sierozems), reddish-brown, lithosols, regosols, solod, solonchak

TABLE 7.6a (contd)

Soil order	Diagnostic properties	Equivalent soils
Mollisols	Dark brown to black, mellow epipedons; one or more diagnostic horizons except spodic or oxic horizons; often a calcareous horizon	Chernozems, prairie (brunizem), chestnut, red prairie, humic gleys, planosols, black rendzinas, some brown forest, reddish-chestnut soils, solonchak, solonetz
Spodosols	Spodic horizon; usually also an albic horizon; usually coniferous forests over sandy soils	Podzols, brown podzolic, groundwater podzols
Alfisols	Pronounced argillic horizon without a spodic or oxic horizon; may have a natric horizon; no pronounced colour changes vertically; high base exchange	Grey-brown podzolic, grey forest, non-calcic brown soils, planosols, half-bogs, solods, terra rossas
Ultisols	Red or yellow argillic horizon; no oxic horizon; low base exchange throughout	Red-yellow podzolic, reddish-brown lateritic, rubruzem, humic gleys, low-humic gleys, groundwater laterites
Oxisols	Oxic horizon, with many free sesquioxides; little horizon differentiation; deep soils	Latosols, ferrallites, groundwater laterites, laterites
Histosols	A surface organic layer at least 30–45 cm thick	Peat, muck

For Table 7.6b *see pp. 138–9.*

SOIL CLASSIFICATION SYSTEMS IN EUROPE

The French system may be taken as representative of the European system of soil classification, although there are various other systems like the Belgian, the English, and the Scottish. The systems concentrate on profile characteristics. Accordingly, hydromorphic, calcimorphic, azonal, and other variants are incorporated in the main system, rather than separated as in Table 7.2 (cf. Table 7.7). The symbols used are as follows:

(A)   simple disaggregation of parent material
A   leaching and/or accumulation of organic matter
(B)   subsoil horizon distinguished on structure only
B   subsoil horizon distinguished on texture

See Table 7.7 *on pages 140–1.*

Apart from greatly easing the problem of soil mapping by doing away with the zonal, intrazonal, and azonal idea, this system also retains the hierarchical arrangement of the soils. The European influence is also great on the FAO/UNESCO (International) system discussed below.

Apart from the French system, the soils of Europe, or part thereof, have been classified, using slightly different systems (cf. Kubiena, 1953; Avery, 1956). Like the French system, Kubiena uses the diagnostic horizon sequence of the total profile, such as ABC soils, (A)C soils, etc. He tries to avoid a system in which selected properties are arranged in order of sequence to establish hierarchical classification, claiming that his classification is a 'natural' one as opposed to the 'artificial' hierarchical systems. This system is shown in Table 7.8. It emphasizes drainage and parent material more than any other factors. Avery's classification of British soils typifies systems designed for a specific purpose—agriculture. The system like others in this category cannot be universally applied, since many of the world's important soils are not represented (cf. Table 7.9, p. 142).

TABLE 7.8
Kubiena's System of Soil Classification

1	Sub-aqueous	1	Soils not forming peat
		2	Peat
2	Semi-terrestrial soil	3	Raw soils
		4	Marsh
		5	High moor peat
		6	Salt soils
		7	Gleys
		8	Alluvial soils
3	Terrestrial soils	9	Climax raw soils
		10	Rankers
		11	Rendzinas
		12	Steppe soils
		13	Altered soils on calcareous rocks
		14	Red and brown loams
		15	Latosols
		16	Brown earths
		17	Pseudogleys
		18	Podzols

TABLE 7.6b
Definitions of Some Diagnostic Horizons of the New American Soil Classification System

Diagnostic horizon	Nomenclature	Definitions
Surface horizons (epipedons)		
Mollic	Ao	Thick, dark surface layer saturated with divalent cations; with low C/N ratio and moderate to strong structure.
Anthropic	Ap	Similar to mollic epipedon in all respects but contains more than 250 ppm acid-soluble phosphate, formed by long-continued farming.
Plaggen	Ap	A man-made surface layer more than 50 cm thick; character varies with the nature of the virgin epipedon from which it was derived.
Histic	Ao	Thin surface layer, saturated with water for part of the year, and high in organic carbon.
Subsurface horizons		
Argillic	Bt	An illuvial horizon enriched with silicate clays translocated there.
Natric	Bt/Bsa	An argillic horizon with prismatic or columnar structure, and saline, having more than 15% exchangeable sodium content.

TABLE 7.6b *(contd)*

Spodic	Bs/Bf/Bh/Bfh	Illuvial accumulation of free sesquioxides accompanied by organic matter; or illuvial accumulation of iron oxides with equal amounts of crystalline clay or illuvial accumulation of organic matter.
Cambic	Bm	A changed or altered horizon, with formation of structure, liberation of iron oxides, formation of silicate clays, or obliteration of most evidence of original rock structure.
Oxic	Bf	Horizon from which has been removed or altered a large part of the silica combined with iron and aluminium, but not necessarily the quartz or 1:1 lattice clays; giving a concentration of clay-size minerals consisting of sesquioxides and 1:1 lattice clays.
Calcic	Bca	Horizon enriched with secondary carbonate, not less than 15 cm thick.
Albic	Ae	Horizon from which clay and free iron oxides have been removed, or in which oxides have been segregated such that the colour of the layer is determined primarily by the sand and silt rather than by the coatings on these particles.

Compiled from Cruickshank, 1972, p. 108.

TABLE 7.7
The French System of Soil Classification

1 Lithosols (sols minéraux bruts)	(A)C	climate	arctic polygonal soils
			desert soils
		non-climatic	erosional lithosols
			depositional lithosols
		climatic	tundra
			alpine
			zero ranker
2 Rankers	AC	non-climatic hydrological	young alluvium
		steppe	chernozem
			prairie
			brown soils
			sierozem
3 Calcimorphic AC soils	AC	calcareous	rendzina
		hydromorphic	brown calcareous soils
			black tropical soils
			calcic alluvial soils
	A(B) or ABC	non-hydromorphic	brown forest soil
			lessivé
		hydromorphic	brown alluvial soil
			lessivé with pseudogley
		non-hydromorphic	podzols
4 Evolved	ABC	hydromorphic	red non-leached

SOIL CLASSIFICATION

5 Ferruginous soils	ABC or A(B)C	red ferruginous soils of tropics (non-hydromorphic)	RFT non-leached RFT leached RFT with ironstone cuirasse RFT with water-table ironstone
		red ferruginous soils of tropics (hydromorphic)	
6 Ferrallitic	ABC or BC	ferrallitic—freely drained	weak ferrallitic normal ferrallitic humic ferrallitic ironstone water-table cuirasses
		ferrallitic—hydromorphic	
7 Halomorphic	AC		saline
	A(B)C or ABC		alkali solonetz solod
8 Hydromorphic	AC	continental	surface pseudogley deep pseudogley gley
		marine	polders
9 Hydromorphic organic			peat-bog half bog

TABLE 7.9
Avery's Classification of British Soils

A Terrestrial soils	1 Raw soils
	2 Montane humus soils
	3 Calcareous soils
	4 Leached mull soils, brown earths, etc.
	5 Podzolized (mor) soils
	6 Alluvial soils
	7 Gley soils
	8 Gley podzolic soils
	9 Peaty soils
	10 Peat soils

SOIL CLASSIFICATION SYSTEMS IN THE UNITED STATES

The Americans, as noted earlier, started with soil classification systems which emphasized the texture of the topsoil as well as the parent material, i.e. the empirical or taxonomic system. Later, as a result of the work of Marbut among others, they shifted to the genetic system, which essentially is the system of the great soil groups. After several decades of use, however, the US Department of Agriculture discovered a number of shortcomings in the great soil groups system.

(1) It does not make room for soils with distinct characteristics, especially those related directly to human utilization. This is a very serious defect in places like Europe and Asia among others, with a long history of human occupance. Vast areas on almost every continent are also presently being urbanized and so are no more natural: scientific agriculture has also greatly changed the nature of most soils.

(2) It does not make provision for multiple soil profiles, such as result from climatic changes. Soil periodicity and the K-cycle concept of soil formation are now widely recognized and accepted, as shown by the literature, especially on Australian soils.

(3) The typical zonal soils are taken as matured soils developed over undulating, well-drained upland surfaces, whereas quite often these are exceptional rather than customary cases.

(4) Many tropical and arid-land soils do not seem to fit the logic of the classification. For instance, some soils combine the properties of podzolization and calcification, which processes could not operate together at the same place according to the logic of the great soil groups.

In short, as more information becomes available, existing systems need to be modified. Accordingly, a working group—the National Co-operative Soil Survey—was chosen in the 1950s to develop a new system to replace the great soil groups. The new scheme should not only allow all US soils to be mapped, but also soils anywhere in the world.

The new US soil classification system is based on the morphological characteristics of the soil profile and, especially, on the recognition of certain 'diagnostic horizons' which reflect the dominant soil-forming processes and environments (Fig. 7.6b). Many of the features of the great soil group classification are, however, retained, e.g. recognition of the stages of profile development and the principles of zonal, intrazonal, and azonal soils. Nevertheless, none of the great soil group names is retained. Rather, the names used have classical language roots in botany and zoology. Ten global orders are recognized, subdivided into suborders, great groups and subgroups, and soil families.

THE UNESCO/FAO SYSTEM OF SOIL CLASSIFICATION

There is no doubt that there is need to harmonize the various systems of soil classification. Accordingly, a number of world bodies, especially those connected with education—UNESCO—and with agriculture—the FAO—have been working in this direction. A number of reports have been issued on the effort to produce an FAO/UNESCO soil map of the world, based on an internationally agreed system. Such a system should perforce be an amalgam of the existing systems, particularly those related to the great soil groups idea and the new American system. When the FAO/UNESCO project is complete, all the continents would be covered at a scale of at least 1 : 5,000,000. This means mapping at the order or, at the best, the suborder level. Larger scales are also envisaged, e.g. the 1 : 1,000,000 soil map of Europe. It is hoped that when this project is completed we shall perhaps have an agreed system of (a) terminology, (b) nomenclatures, and (c) taxonomy.

Like many other attempts to harmonize widely diverse systems, the result of this international effort is likely to be a rather unwieldy system. Thus, by 1964, as many as forty-one soil units were being proposed.

Many of them are yet to be agreed upon while others are certain to be proposed. Examples of continental soil maps that have so far been produced—e.g. the soil map of Africa—illustrate very adequately the problem faced by international organizations.

The Soil Map of Africa

Mapping the soils of Africa presents particular problems and so needs to be mentioned at this stage. Besides the problem of inadequate survey over large areas, the systems in use vary widely from place to place, mainly according to those used by the former colonial masters. The various political divisions fall into three main groups: (a) the French-speaking, (b) the English-speaking, and (c) the others, including Portuguese and Spanish colonies. The systems of soil classification used in each of these groups differ in detail, but are not irreconcilable. For instance, both the English-speaking and the French-speaking systems are based on soil genesis and so reflect the influence of climate, particularly rainfall. These various systems have been described already. Nevertheless, there are areas of differences which make the task of integrating them very difficult.

The task of integrating these various systems to produce a comprehensive map fell on a group called the Inter-African Pedological Service (SPI). It was started in 1953, but the decision to produce a map on the scale of 1 : 5,000,000 in collaboration with the four Regional Committees for the Conservation and Utilization of the Soil was actually taken in November 1955. The mapping process took almost ten years, as the final map (D'Hoore, 1964) was published in 1964, after it had been approved at a meeting of the Commission for Technical Co-operation in Africa (CCTA) in 1963. It is therefore sometimes called the CCTA map. The map is an amalgam of the various mapping and classification systems as well as of all available soil maps and extrapolations. An assessment of the reliability of the map shows that about 50 per cent of the area was covered by partial maps of good quality; 30 per cent was based on less precise maps; and 20 per cent was extrapolated on the basis of some hypotheses and generalization.

The map is published on eight sheets, with a sheet covering West Africa. The legend is summarized in Table 7.10, p. 146. Several of the units shown can be correlated with a number of other systems, particularly the genetic systems such as the Russian system. They can also be correlated with the new American system—the Seventh Approximation—with little difficulty. Most lithosols may be classified either as entisol (lithic soil), as inceptisol or as aridisol; while the subdesert soils may be classified as aridisols. Calcimorphic soils are in part mollisols; vertisols are almost identical with those of the corresponding order in the American system; podzolic soils are spodosols; high veld pseudo-podzolic soils are alfisols, etc. (cf. Table 7.6).

However, correlation becomes more difficult, especially with the ferruginous tropical soils, ferrisols, and ferrallitic soils. The nearest equivalents are alfisols, ultisols, and entisols for the ferruginous tropical soils and oxisols for the ferrisols and ferrallitic soils. It is important that most of these terms apply to soils at different levels of the hierarchy, thus increasing the problem of integration. Nevertheless, the soil map of Africa represents a successful attempt to bring together the various classifications used in Africa into a single system adapted to African conditions. It is therefore of considerable interest to students of African soils. It has accordingly influenced current soil survey and soil-mapping exercises all over the continent.

The system adopted for this map is somewhat different from that being considered by the FAO/UNESCO Commission. It may therefore be expected that the system used on this map will be modified, or another map produced to conform with those of the other continents. On the contrary, this map may influence the other maps which are still being drawn.

See Table 7.10 *on pages 146–7.*

Soil Classification for Specific Purposes

The systems of soil classification discussed so far are mostly for general rather than for specific local applications. The genetic systems as well as those based on important profile characteristics reflect to some extent the agricultural value of the soil. They may therefore be used, in lieu of better maps, to obtain a broad picture of soils in an area in relation to their possible uses. However, for precise information at the local level, such as can provide a basis for precise recommendations for, say, agriculture, additional and/or alternative classifications are needed.

Such classifications are often based on a characteristic or group of characteristics considered important for a particular project. For instance, a map of a farm plot or project area showing the texture of the surface soils and their drainage characteristics is certain to be more meaningful and useful to the farmer than one showing oxisols, alfisols, ultisols, etc. Moreover, soil survey, say, for purposes of irrigation pays particular attention to relief, soil texture, water sources, supply, and quality, and especially the sodium chloride content of the water. Soil depth is also important in terms of a project concerning tree crops. This is particularly important where ironstone or indurated layer occurs in the soil profile.

The special purposes for which soils are classified include land capability and/or potentiality and soil classification for irrigation agriculture. Land capability classification is adequately covered in Chapter 5; irrigation is discussed briefly in what follows.

TABLE 7.10
Legend of 1 : 5,000,000 Soil Map of Africa

Raw mineral soils	
Major soil types	Major morphological features and landscape relationships

Rock and rock debris
    Rocks rich in ferromagnesian
    minerals
    Ferruginous crusts
 	
Desert detritus
    Sands (ergs)
    Clay plains
    Desert pavements, residual
	Calcareous crusts
	Not differentiated
	
	Desert pavements, transported
	Not differentiated

Weakly developed soils	

Lithosols (skeletal soils and lithic soils)
    On lava
    On rocks rich in ferromagnesian
    minerals
Subdesert soils
Weakly developed soils on loose sediments not recently deposited
Juvenile soils on recent deposits
    On volcanic ash
    On riverine and lacustrine alluvium
On fluviomarine alluvium (mangrove swamps)
	On ferruginous crusts
	On calcareous crusts
	Not differentiated
	
	On wind-borne sands
	Soils of caves

Calcimorphic soils	
Rendzinas, brown calcareous soils Soils with calcareous pans	Soils containing more than 15% gypsum
Vertisols and similar soils	
Derived from rocks rich in ferromagnesian minerals Derived from calcareous rocks	Of topographic depressions, not differentiated
Podzolic soils High Veld Pseudopodzolic soils Brown and reddish-brown soils of arid and semi-arid regions	
Brown soils of arid and semi-arid tropical regions	
On loose sediments	Not differentiated
Brown and reddish-brown soils of arid and semi-arid Mediterranean regions	

TABLE 7.10 (contd)

Eutrophic brown soils of tropical regions	
On volcanic ash On rocks rich in ferromagnesian minerals	On alluvial deposits Not differentiated
Red-brown Mediterranean soils	
Red Mediterranean soils	Brown Mediterranean soils
Ferruginous tropical soils	
On sandy parent materials On rocks rich in ferromagnesian minerals	Not differentiated
Ferrisols	
Humid On rocks rich in ferromagnesian minerals	Not differentiated
Ferrallitic soils	

Dominant colour: yellowish-brown
  On loose sandy sediments
  On more or less clayey sediments
	Not differentiated

Dominant colour: red
  On loose sandy sediments
  On rocks rich in ferromagnesian minerals
	Not differentiated

Humid ferrallitic soils Ferrallitic soils with dark horizons Yellow and red ferrallitic soils on various parent materials	
Halomorphic soils	
Solonetz, solodized-solonetz Saline soils, alkali soils, and saline alkali soils	Soils of sebkhas and chotts Soils of lunettes Not differentiated
Hydromorphic soils	
Mineral hydromorphic soils	Organic hydromorphic soils
Organic soils, non-hydromorphic	
Organic soils of mountains	
Associations and complexes	
Complex of volcanic islands	

Soil Classification for Irrigation Agriculture
This system is even more specific than the land capability one. Here, soils are often considered in terms of limitations to successful irrigated agriculture. As in land capability, the system that is widely used is that of the US Bureau of Land Reclamation. The system recognizes six categories or classes. Classes 1 to 4 are all suited to irrigated agriculture, with Class 1 land having the least limitations and Class 4 the most. Class 5 may be brought under irrigation, with some effort, i.e. it is marginal; but Class 6 cannot. The factors of the landscape that are usually considered in this connection are:

(a) The gradient of the land; only flat or gently sloping land is normally considered.
(b) The texture of the soil. The best soil in this connection is loam. Very sandy and very clayey soils are not particularly suited to irrigation, the former because they store so little water and have excessive permeability, the latter because they are relatively impermeable, so that water flows very slowly through them, if at all.
(c) The level or degree of salinity, particularly the sodium chloride content of the soil solution, which controls the degree of toxicity during use.

The following are fairly detailed descriptions of soil classes for irrigated agriculture:

Class 1 Highly suitable. Can produce sustained and relatively good yields at moderate cost. Are flat or nearly so and have little erosion hazard. Are deep; of medium to moderately fine texture; have a structure and porosity allowing easy penetration of roots, air, and water; combine good drainage with an adequate moisture storage capacity. No artificial drainage works are normally called for. Are free of salt or other toxic substances.

Class 2 Well-suited to irrigation agriculture but somewhat more limited than Class 1 soils. May be more expensive to prepare than Class 1 soils; also more costly to farm because they require more supervision. Limiting factors may include one or more of: shallow depth; too coarse (light) or too heavy texture; low moisture-holding capacity, slow permeability rate due to compaction. May require some artificial drainage or reclamation work.

Class 3 Have many more limitations still. May require heavy expenditure in land preparation, more care during use, greater use of fertilizers and treatments, etc. May be so restricted by heavy texture and poor drainage that only partial correction may be possible, and that at high cost.

Class 4 Are severely restricted in their irrigation possibilities. Suited only to special types of land use, e.g. pasture, otherwise require heavy investment in engineering and drainage works before they can be used. Factors that may account for this condition include: very poor drainage because of heavy texture; unfavourable low position causing periodic flooding; very high water table; steep valleyside slope; etc.

Class 5 Marginal; can be brought under irrigation under certain circumstances. For example, land may be so scarce as to justify heavy engineering works, e.g. extensive reclamation of saline (toxic) soils; use of powered pumps; etc. Great necessity warrants such efforts.

Class 6 Non-irrigable under any circumstance. Cannot be reclaimed, or otherwise modified, to justify irrigation. Examples are: rugged, steep-sloping country; extreme desert areas with no source of water anywhere nearby; extremely saline (toxic) conditions that cannot be redeemed; very heavy or essentially gravelly soils; etc.

Irrigation suitability maps are not yet available in large parts of Africa. What we have are areas where irrigation agriculture is in progress or planned. Irrigation suitability maps are, however, very useful for planning purposes, especially for deciding whether an irrigation scheme warrants the envisaged investment. However, such maps only provide general guidelines; detailed study is essential for each irrigation project. What the map does is to indicate the type of limitations involved by adding suffixes to the irrigation class; e.g. S for poor or unsuitable soil characteristics; T for rough topography and/or steep slopes (too high gradients); D for poor drainage conditions; R for loose rock (gravel and bigger) particles within the plough layer; and C for climatic limitation, other than low rainfall.

A Suggested System for Soil Classification

The geographical study of soils of necessity involves a global coverage. This is a very difficult task, given its infinite and complex variabilities in space and time. Indeed, it is this fact, as observed earlier in this chapter, which has prompted the idea of classification in the first place. It has also resulted in the different systems now available in the literature, only a few of which are mentioned here. Moreover, some of these systems are too involved, for either routine classroom teaching or for soil survey and mapping.

TABLE 7.11
Simple Soil-forming Processes

Process	Weathering products					
	Alkali cations (Na$^+$, K$^+$)	Alkaline earth cations (Ca^{++})	Sesquioxides (iron and aluminium)	Silicate clays	Humic acid humus	Fulvic acid humus
1 Removal from soil profile	—	—				—
2 Accumulation throughout the profile		—	—	—	—	
3 Illuviation to B horizons	—	—				
4 Capillary rise to upper horizons		—	—	—		

— means the process occurs; a blank space means the process is rare or unlikely
Modified from Corbett, 1969.

The problem is to evolve a system which will not only reflect the important properties and characteristics of soils as natural bodies but which is also simple, logical, and easy to understand. This is perhaps why the morphological approaches to soil classification have been popular all along. The most successful of these are those which also reflect the formative factors and processes, i.e. combining the generic and the genetic systems. The new American system and some Australian systems are useful examples of these. The former also has the advantage of including man as a soil-'forming' agent but, as it now stands, it is too cumbersome to be useful, especially in teaching. The same thing is true of the genetic (great soil group) system.

It is possible, from a study of soil profiles in relation to the weathering and pedogenic processes of their formation, to evolve a less complex, less cumbersome system. We can, for instance recognize six products of weathering and pedogenesis. The products of mineral weathering most conspicuous in soil properties are (i) the alkali cations (sodium, potassium, etc.); (ii) the alkaline earth cations (calcium); (iii) iron and aluminium cations; and (iv) the silicate clays. In addition, we have two main types of humus, namely, (v) humus which is high in mobile, fulvic acid component, or brown humus; and (vi) humus which is high in the less mobile, highly coloured, humic acid component or black humus. We can also recognize at least four things which can happen to each of these products, depending upon the conditioning influence of the soil-forming factors. These are: (i) complete removal from the soil profile, e.g. where leaching is intense and so predominates over accumulation; (ii) accumulation throughout the soil profile, e.g. where the rate of formation is faster than that of leaching or loss; (iii) illuviation from the A or E to the B horizons, where they accumulate; and (iv) accumulation in the upper soil horizons, due to capillary rise and precipitation. This therefore means that, at least in theory, there are twenty-four possible processes of soil formation (Table 7.11).

However, not all the twenty-four combinations or processes occur in nature. The removal of silicate clays from the profile is not likely, since they move in suspension rather than in solution. Similarly, iron and aluminium are not very soluble, and once they are deposited as sesquioxides, they are insoluble and so are unlikely to be completely removed from the soil profile. Sesquioxides also move as ions in solutions and not as particles as in silicate clays; they can therefore move freely upwards and downwards through the soil's capillaries. Upward movement of clay is uncommon, mainly on account of the larger size of colloids which then must move in suspension rather than in solution. Furthermore, clay particles tend to block the soil's capillaries when moving up and so prevent further upward movement.

TABLE 7.12
Relationship Between Soil Groups Used in This Book and the Other Popular Systems of Soil Classification

	This book	Corbett, 1969	American 7th approximation	Great soils groups	French system	CCTA soil map of Africa
1	humid tropical soils	soils dominated by sesquioxides	oxisols, ultisols	laterites (latosols), lateritic soils, terra rossa	sols ferrallitiques	ferrisols, ferrallitic soils, ferruginous tropical soils
2	podzolized soils	podzolized soils	spodosols, alfisols, ultisols	podzols, brown podzolic, groundwater podzol, brown forest podzolic, red podzolic, yellow podzolic, grey-brown podzolic, grey podzolic	podzols, brown forest soils, sol lessivé (evolved mull and mor soils)	podzolic soils, red and brown Mediterranean soils
3	soils rich in bases and silicates	(1) soils dominated by silicate clays (2) soils dominated by	mollisols	chernozems, chocolate soils, prairie soils, black earth, terra rossa,	calcimorphic soils	calcimorphic soils, vertisols

4	gleyed soils including recent soils	soils dominated by organic matter	sand sierozem	inceptisols, histosols, entisols	solonchak, solonetz, solodized-solonetz, solodic desert loam, desert	alpine humus soil, acid moor peat, humic gley soil, fen soil, moor podzol peat, azonal soils	hydromorphic soils, continental, marine	hydromorphic soils
5	relict (fossil) soils	relict soils of Australia				lateritic krasnozem, lateritic red earth, calcareous lateritic, podzolic hardpan soil	ferruginous tropical soils, ferrallitic soils	ferruginous tropical soils, ferrisols, ferrallitic soils

Similar limitations apply to the soil organic matter. Humus rich in humic acid is generally immobile, and so is unlikely to be illuviated or to rise through the profile. By contrast, humus high in fulvic acid is mobile and so does not generally accumulate in the topsoil, i.e. it tends to be illuviated. Once it is deposited in the B horizon, it becomes fairly insoluble, since it does not rise in the profile.

Finally, the weathering products have different rates and degrees of solubility and this explains the presence or absence of the various substances in the soil body. Sesquioxides are mostly insoluble, but sodium chloride is very soluble. Salt accumulation is unlikely to operate together with eluviation or capillary rise of iron and aluminium. This is because the former requires very weak and the latter very strong (intense) leaching. Similarly, accumulation of the bases does not occur together with the eluviation of humus or iron because the latter requires rather low pH values.

Consequently, there are only a few combinations of the simple processes occurring in nature, as shown in Chapter 4. The major ones are (1) lateritization, which denotes the accumulation of sesquioxides at the expense of all other constituents or weathering products; (2) podzolization denoting base depletion, surface accumulation of organic matter, eluviation of humus, iron, or clay; (3) solodization denoting salt accumulation accompanied by some degree of podzolization; and (4) gleization denoting poor drainage and the formation of gleys and mottled soils. These processes operate in distinct climatic regimes and produce soils which are different from one another, whether in the component substances or in terms of soil morphology. In other words, it is possible to classify soils according to these major soil processes. Lateritization may combine processes such as ferralization, ferruginization, etc., while calcification and salinization may be grouped with solodization. Podzolization may form a major category and so may gleization.

At the highest (or order) level of soil classification, therefore, we may have: (1) laterites or soils dominated by sesquioxides; (2) podzols and podzolized soils characterized by surface organic matter accumulation, base depletion, and eluviation of humus, clay, and iron; (3) base-rich soils of arid, semi-arid, and subhumid regions; and (4) soils of the periglacial and glacial regions, e.g. gleyed soils. These are soils found respectively in the tropical, temperate, dry, and cold regions, the four major terrestrial ecosystems commonly recognized on the earth's surface. These soils may be further distinguished on the basis of the intensity and also the details of the processes involved. Possible basis for such division may include climatic or vegetation zones, or parent material. Further subdivisions on basis of topography, drainage, etc. are also possible. Finally, due to climatic changes among other things, some soils now found in different parts of the world are not directly related to present-day processes but were formed long ago under different pedogenic regimes. These soils may be a major category in which case they cut across the other categories (Table 7.12). Otherwise they may form subgroups wherever they occur in the above-described scheme. The best example of these fossil or relict soils are the duricrusts—laterites, ferricretes, and silcretes—which are widespread the world over, irrespective of latitudinal location. This is the system we shall adopt in the next chapter to describe the world soils. It is quite similar to the system used by Corbett (1969) and Bridges (1970).

Chapter 8

WORLD SOILS

A global coverage of soils necessarily involves a high degree of generalization. Accordingly, world soils are commonly described at the 'great soil group' level, at which level about nine 'principal zonal soil groups' are usually considered. Apart from these, intrazonal and azonal soils are also included. Currently, great soil groups seem to be giving ground to the new American (Seventh Approximation) system (Fig. 8.1). This latter system maps eleven 'soil orders', the boundaries of many of which coincide (or almost do so) with those of the great soil group system.

Apart from these two systems, world soils have been mapped and described in a number of other ways (see Chapter 7). As hinted in the last chapter, the level of generalization envisaged in the present work is much higher than that used by the two systems described above. The six major categories identified in that chapter will now be described.

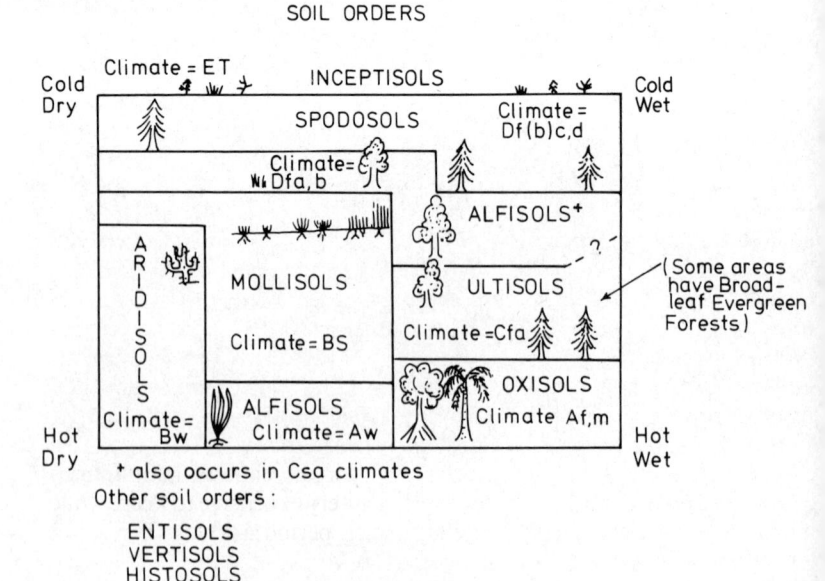

Fig. 8.1   New American comprehensive soil classification system

The Soils of the Humid Tropics

This is one of the most written-about soil categories in the literature, but very little about it is yet known for certain. The soils have been called many names—laterites, lateritic soils, latosols, and oxisols. Indeed, the discussions and controversies over the precise nature of these soils, how to define them, and what to call them make up a substantial part of the literature on the topic.

Like most zonal soils, the descriptions in the literature rest greatly on theoretical (conceptual) grounds rather than on real field experience and documentation. This is to say that very few convincing studies have emerged to support the generalization one finds in the literature on these soils. Studies exist for particular areas—e.g. South-East Asia, Brazil, Central Africa—but we are yet to have a clear picture of these soils, comparable, for instance, with that of the podzols and podzolic soils.

The reasons for this situation are many and varied. First, most of the books available at the present time are written by people from extra-tropical areas. Some of these authors never visited the area at all, while others made only short visits. Most authors therefore have relied on information obtained from mostly unreliable secondary sources, particularly those based on the assumed relationships between climatic parameters, soil formation, and soil types.

Second, even for those who stayed in the area for a fairly long time, or actually live there, there are a number of predicaments which affect the results of their studies. Among these are: (a) the initial bias due to their foreknowledge of the soil situations in the temperate regions. This is particularly serious because almost all the soil scientists who have worked in the tropical world so far have been trained in the temperate lands and so are familiar first with the soils and soil-forming processes in those areas. The result, except where this factor is recognized and guarded against, is the application of the concepts of temperate soils and soil formation to tropical soils. (b) There are also the problems associated especially with the tropical climate and vegetation. The high temperatures, humidities, and rainfall and the accompanying dense, sometimes impenetrable vegetation make fieldwork very difficult, if not impossible. The soil survey system, therefore, has to be modified accordingly. (c) Another problem is that of inadequate and unsuitable personnel, whether to organize soil survey or to assist the organizers of such survey. Most countries in the tropical world depend on foreign experts, mainly from Europe or North America. These experts are very expensive to keep, while many are even not willing to stay for a long period anyway. The result is a number of unco-ordinated *ad hoc* studies, largely restricted to 'project' areas as observed in Chapter 5. (d) There is also the fundamental problem of the humid tropical pedogenic environment. This, among other things, tends to make weathering depths and intensities vary widely within short distances, because of such factors as micro-climate, geological structure, relief, etc. (e) The pedogenic processes operating in the tropical environment are also not properly understood. Unlike, again, podzolization which has, among other things, been reproduced experimentally in the laboratory, lateritization has not been so tested. It is in this connection that Pendleton (quoted by Reiche, 1962, p. 71) once remarked: 'I should add that it will be quite impossible for you to get a clear picture of laterites and lateritic soils from the literature.' (f) Tropical soils are among the most complex and so are by their very nature very difficult to study. Not only is the profile often poorly developed, compared with, say, the podzol profile, the depth also varies very widely within very short distances, while it is nearly impossible to locate precisely the boundary between soil *per se* and the soil parent material. It is therefore not surprising that there is no uniformity or standardization in the study of tropical soils in the different countries. This situation is made worse by the fact that different European nations with their various approaches initiated soil study in different areas of the tropical world. The Belgians, the Dutch, the French, and the English, with their different backgrounds and systems, studied soils in the areas under their control. It is only very recently that attempts have been made to bring the various sources together.

Following the First Commonwealth Conference on Tropical and Subtropical soils held in England in 1948, many international conferences have been held during which attempts have been made to clarify the situation with regard to the information available on tropical soils. Many institutions have also been established, e.g. the International Institute of Tropical Agriculture (IITA), concerned essentially with the study and use of tropical soils. Besides, a number of international consultants have carried out regional soil surveys and land classification projects (e.g. the Land Resources Division of the British Overseas Development Ministry; Chapter 5). Yet not much progress has been made. Many misconceptions, ill-defined terms and concepts, and incompatible systems still exist. The problems, for instance, of studying tropical soils by concepts, methods, and techniques based on information acquired in cool and temperate regions, rather than by those adapted to local conditions, still rear their head. This is, of course, an example of a particular problem which, as recently realized, cannot be solved by the general approach. Rather, it requires specific approaches, methods, and techniques. On the whole, these have not been developed for tropical soils. Our knowledge about the soils is not only insufficient but it is also confused.

Nevertheless, we shall in what follows try to present these soils in as simple and comprehensible a way as possible, based on a reconsideration of the information available from the literature, as well as on field experience. First, we try to identify some common or unifying influences or factors which govern the soil-forming processes. Next we review briefly the information available in the literature on these soils. Finally, we describe the authors' work and experience on these soils. A similar procedure will be followed with respect to the other soil categories.

The Humid Tropical Environment and Soil Formation

The climate and associated biotic life of an area are obviously the most significant unifying factors in soil formation. Being the major soil formers, variations in them are reflected in the soil-forming processes, and consequently in the soils of an area.

The humid tropical zone has been variously defined. However, most people take it as corresponding roughly to the area covered by Koppen's A climate (Fig. 8.2), a zone that straddles the equator but tends to be somewhat wider on the eastern than on the western side of the continents. This area is characterized by:

(a) Hot wet conditions, with the coolest monthly temperature never below 18°C. This means that there is no winter season. There is also a generally low temperature range, both diurnal and annual.
(b) Rainfall is sufficient for tree growth, i.e. it is hardly less than about 750 mm per annum.
(c) Three major subtypes, based mainly on rainfall characteristics, are known. The Af (Equatorial) type has continuous rainfall with no month having less than 60 mm. The Am (Monsoon) type has a seasonal distribution but its annual total is high enough to offset the dry spell. Thus Af and Am types have a rainforest type of vegetation. Finally, the Aw (Sudan) type has light and seasonal rainfall and so is characterized by dry forests and savannas. Apart from rainfall, temperature also distinguishes these types. In short, the three major types constitute major pedogenic provinces in which slightly different soil-forming processes operate. It is possible to subdivide each of these but such a detail is not needed in the present work.

The most important aspects of the soil-forming environment are not, however, the atmospheric conditions as described above, although they

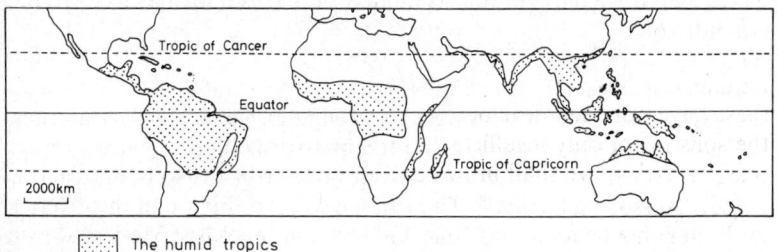

Fig. 8.2   The humid tropics

influence it indirectly. The important factors are the soil climate and soil biology, which together control the rate of the processes of soil formation including weathering, leaching, humification, mineralization, etc. The main feature of the pedoclimate in the humid tropics is the abundance of heat and moisture, particularly the former, all the year round. The soil moisture in this region is always warm, averaging 24–25°C. This is about 15·5°C warmer than is, say, northern Europe or Canada. The soil body is therefore quite warm all the year round.

The result can be summarized briefly as follows:

(a) The ionizing power of the soil water is quite high, about four times as high as at a temperature of 10°C.
(b) Silica is several times more soluble than in the cool environments of the temperate and cold lands.
(c) The weathering process of solution in particular is greatly accentuated. This means there are more elements or ions in solution passing through the soil body. The result is higher hydrolytic influence of such solutions. In short, rocks weather to greater depths in this than in any other environment, thus providing opportunity for the development of deep and old (mature) soils.

Like the tropical pedoclimate, the biotic environment in the humid tropical regions is very conducive to rapid soil formation. The plant and animal life produces plenty of organic matter, which is also decomposed very rapidly. The process of mineralization of organic matter is the most rapid on the globe. Where soil water is plentiful and temperatures are low, this process may be hampered and organic matter may remain undecomposed or partly decomposed, as in the case of peat soils. This situation is not typical of the humid tropical environment.

In short, the humid tropical environment is such that it is suitable for the development of the deepest profiles both of weathered rock and of soils. All other things being equal, soils in the humid tropical regions will develop to the most advanced stages. Finally, the combination of tropical climate and vegetation produces a situation which maximizes leaching processes but encourages the accumulation of sesquioxides to form the *oxic* horizon.

Examples of Humid Tropical Soils

The soils of the humid tropics were for a long time generalized as laterites. This is certainly not the case. In the attempt to dispel such a generalization numerous other terms have been coined to describe the soils. Such terms include 'latosol' and 'oxisol'. The controversy on the use of these terms has been going on for a long time and still continues (see Mohr and Van Baren, 1959, pp. 353–6). Suggestions have even been made to dispense entirely with the terms 'laterite' and 'lateritic' soil for these other terms, but these have not succeeded; the terms are still being used and rightly so (Faniran, 1969).

The reason for this is simple. The term 'laterite' was the first to be applied to the material in question, and so has achieved great popularity. Coined from the Latin *later, lateris*, meaning a brick, the term 'laterite' was first used by the English explorer and traveller Francis Buchanan (later Lord Hamilton) during his journey through Mysore, Canara, and Malabar (in India) in 1807. It was applied to the material he found at Angadipuran, which he observed to harden on exposure to the air and which was used extensively by the local people for building. Buchanan described the material as '... immense masses, without stratification, full of cavities and pores ... containing a very large quantity of iron in the form of red and yellow ochres'.

Since Buchanan's time, this term has been widely misused. Indeed it is perhaps the most misused of all pedological terms today. It has been applied not only to a soil profile but also to a wide range of materials including silcretes, ferricretes, bauxites, calcretes, etc. It is therefore not surprising that people got tired of it. It is necessary to define this term specifically, if we are to continue to use it.

The following clarifications are offered. First, laterite refers to a particular material or portion of a soil profile. This material is usually concretionary, nodular (pisolitic), vesicular, vermicular, or massive, depending on the nature of the rock from which it has formed. Pisolitic duricrusts tend to form in sandstones and similarly coarse-grained rocks while fine-grained rocks tend to make essentially massive, vermicular, or vesicular duricrusts. Whatever its physical characteristics, the duricrust is usually associated with the indurated zone of a soil (deep-weathering) profile, which is typically underlain by the mottled zone and/or the pallid zone, overlying weathered, weathering, or unweathered country rock. The term 'duricrust' has, however, also been applied to all surficial crusts, whether or not they are connected with deep weathering (cf. Dury, 1969).

The mineralogy of laterites is also of interest. The most important minerals are the hydroxides of iron and aluminium, which are usually present in substantial quantities. The material is called bauxite if it is essentially aluminous, and iron oxide or ferruginous duricrust if it contains essentially iron hydroxide. Other minerals commonly found in laterites are titanium oxide and manganese oxide. Titanium is commonly present in small quantities only, mostly less than 5 per cent. By contrast, manganese oxide, like iron and aluminium, may make up the essential part of the duricrust, in which case it is called manganese duricrust. Finally kaolinite and silica may be present in substantial quantities in all the types of duricrust, including laterite (Grubb, 1965; Faniran, 1970).

The ratios of these minerals or elements have been used to define them. However, such information is rather unreliable, as shown by analyses

TABLE 8.1
Chemical Analysis of Duricrusts and Other Weathered Products from the Sydney Area

Sample parent material	Chemical content, wt % chemical ratio					
	SiO_2	Fe_2O_3	Al_2O_3	TiO_2	MnO	SiO_2/sequioxide ratio
1 Dolerite	5·90	48·20	41·50	1·40	0·60	0·06
2 Shale (Wianamata)	35·60	40·50	22·90	0·80	0·20	0·55
3 Shale (Narrabeen)	24·50	51·10	23·70	1·00	—	0·32
4 Fine-grained sandstone	24·40	48·30	25·30	1·10	—	0·32
5 Fine-grained sandstone	19·30	24·70	53·30	0·20	0·60	0·24
6 Coarse-grained sandstone	39·80	23·80	35·30	0·10	0·11	0·64
7 Coarse-grained sandstone	48·90	17·30	32·80	0·90	0·20	0·95
8 Conglomerate	72·10	17·90	9·30	0·56	0·05	2·58

done from many parts of the world. Analyses of the Sydney laterites, for instance, show that the silica/sesquioxides ratio may be anything between 1 and 10 (see Table 8.1).

Finally, the same types of minerals occur in the underlying zones, particularly the mottled zone, but the amounts and proportions vary. Iron, aluminium and the other sesquioxides tend to decrease in amount while silica and kaolin tend to increase with depth in the soil profile (see Table 8.2).

TABLE 8.2
Polished Section Modal Analyses of Laterites in Fine-Grained Sandstone,* Sydney District

Sample reference		Mineral content (wt %)		
		Maghemite	*Other opaques*	*Rest (mainly quartz)*
Upper-indurated zone	1	33.9	53.7	13.0
	2	28.7	62.2	9.1
Middle-indurated zone	3	16.2	69.0	14.8
	4	4.6	81.4	14.0
Lower-indurated zone	5	3.1	77.8	19.1
	6	1.2	84.4	14.4
	7	0.9	90.0	9.1
	8	0.0	53.7	46.3

* Similar trends were observed in respect to laterites formed in shale and in coarse-grained sandstone.

The second point to note about *laterite* is that its profile typifies the deep weathering (duricrust) profile (Chapter 4). The same number of zones—indurated, mottled, and pallid—occur in both. By implication, therefore, the same environmental conditions are likely to apply.

The nearest equivalent, in the literature on soils, of the laterite profile as defined, is the ferruginous tropical soil profile (sols ferrugineux tropicale), associated with the savanna or other open forest vegetation. The other soils—e.g. the ferrallitic soil, the ferrisol, and the vertisol—although they may show stratified phenomena, are not strictly speaking laterites, except where they develop a hard-crust layer at the top, presumably following a clearing of vegetation, a change in climate, or other changes that may affect the groundwater situation. The laterite profile is, therefore, only one example of deep-weathering profiles found in the humid tropical regions, and so cannot be generalized over the entire region.

The third point about the laterite profile is that it is often quite old.

It is, in fact, fossil in most areas where it occurs, being associated with old erosion surfaces, particularly the Tertiary and pre-Tertiary surfaces. This means that the profile should have been widely destroyed and the material widely redistributed. Such materials have also considerably aided the present-day weathering and pedogenic processes, since they provide abundant amounts of sesquioxides. The deposited materials have therefore been reworked in many places into typical laterite profiles. The materials so formed are, however, different from the primary laterite. The indurated zone is usually more heterogeneous, less compact, and has quartz pebbles embedded in it. They have therefore been called secondary or reworked laterites. Reworking is particularly common where rainfall and temperature are adequate, as in the humid tropical plains. Similar materials have also been called 'cemented pediment gravel' or the 'Kongi/Agodi gravel' in Nigeria (Burke and Durotoye, 1970).

Finally, because of the variations which characterize the humid tropical environment, there are bound to be variants of the lateritization process, the term 'lateritization' being applied generally to the process in which silica is leached from the soil profile alongside the bases, and in which the sesquioxides are concentrated. The process also includes dehydration and hardening of the top (indurated) zone.

Accordingly, many types of soil have been recognized and described in the literature. Examples of soils developed in this environment are laterites, lateritic soils, ferruginous tropical soils, red loams, yellow loams, vertisols, and krasnozems.

However, the soils can be grouped, on morphological grounds, into two classes, namely, those which show definite zones or stratification, and those in which stratification is poorly developed or not at all which are referred to as krasnozems. The stratified soil profile may or may not have a laterite layer. If it does it is a laterite profile, otherwise it may be described as lateritic. The soils of the humid tropical regions, especially those which are fully developed, may therefore be described as laterites, lateritic soils, and krasnozems, with the proviso that all possible gradations between them are recognized. In other words the three soil classes or soil categories are members of a continuum, while local factors of relief, drainage, bioclimate, vegetation, dissection, land use, etc. further affect the pedogenic processes and so further complicate the simple situation or picture here presented.

LATERITES OR DURICRUST-TYPE (HARDPAN) SOILS
Laterites as described above have well-defined, well-developed zones or horizons, including the indurated zone underlain by the mottled zone and the pallid zone. The conditions necessary for lateritization or duricrusting have been widely discussed in the literature. Some of them are

still controversial, but suitable rock type or ready sources of sesquioxides, low relief, periodic wet and dry seasons, open vegetation, and a long period of stability are generally considered to be essential conditions. Laterites are therefore best developed in the dry forest and savanna areas of the humid tropics, where the seasonal rainfall pattern, the extensive plain surface, and the open vegetation seem to have aided the process of sesquioxide accumulation and the formation of the indurated zone. By contrast, the hardpan is absent in many high forest continuous rainfall (Af) areas. However, clearing and periodic desiccation have resulted in the formation of the hardpan even in the forest regions. Most forest soils in this region are therefore potentially lateritic soils and so may be described as such, especially when formed under the influence of the groundwater table.

The laterite profile typically consists of the following:

I A thin humus layer, consisting of a litter which is being decomposed and mineralized. The rate of humus decomposition and mineralization is usually very rapid, leading to the non-accumulation of humus. This characteristic, which is shared by all humid tropical soils, makes them very vulnerable to depletion and destruction immediately after clearing. This top layer is the organic soil.

II A layer of soil of variable depth equivalent to the A horizon. This zone varies from a few centimetres to several decimetres or even metres. Unlike in the podzol profile (see below), the effect of leaching is not obvious, except in the clay content. The zone is usually red or brown and may contain pisolites or spherulites, the proportion of which increases with depth to the indurated-zone layer below. The texture is generally light, since the clay has been mostly eluviated to the horizon below.

III The laterite horizon or the indurated zone. This zone in the ideal profile is commonly pisolitic, and comprises concretionary, vermicular, and vesicular material, in different stages of induration or hardening. Studies made in parts of Australia (Wright, 1963; Hays, 1967; Faniran, 1970) show that this zone, where it is best developed, can be subdivided into at least three parts. The topmost part attains the most advanced stage of dehydration and induration. Moreover, the pisolites are smallest, hardest, most spherical, and most plentiful, decreasing in number, hardness, sphericity, and degree of perfection downward through the profile. The associated minerals include haematite (α-Fe_2O_3), maghemite (γ-Fe_2O_3), and gibbsite (Al_2O_3), most of which decrease in quantity with depth. Maghemite in particular is almost invariably restricted to the topmost parts of this layer (Faniran, 1970). The upper indurated zone, as this zone is commonly called, merges, sometimes imperceptibly through the middle indurated zone, to the lower indurated zone. Generally speaking the materials in the lower portions of the indurated zone are less compact, softer, redder, and mostly hydrated, as opposed to the hard, compact, and mostly dehydrated material of the upper indurated zone. The total depth of the entire indurated zone as well as of parts thereof varies very widely. Measurements made, again in the Sydney district of Australia, show that they vary between 30 cm and 122 cm. The profiles studied have also been widely truncated in many places and the materials widely redistributed and reworked (Faniran, 1969). Similar observations have been made in parts of West Africa, particularly in Nigeria, as is shown in what follows. The lower indurated zone finally merges, sometimes through a transitional zone, into the mottled zone.

IV The mottled zone. Like the indurated zone, this zone varies from rock to rock. In the sandstones of the Sydney district, the mottled zone may be essentially iron-enriched quartz, sometimes with mottling. In such cases iron seems to have replaced the original kaolinite, silica, and other cement, while the mottled zones in shales, claystones, and other fine-textured rocks are essentially ferruginous and aluminous clay. The mottled zone is itself sometimes divisible into upper and lower portions. In many situations, the upper part is uniformly red, without mottling, while the lower portion is typically mottled. The upper mottled zone in many profiles represents the transitional zone between the pisolitic indurated zone and the real (kaolinitic) mottled zone. The mottled zone is made up of essentially hydrated minerals, e.g. goethite, lepidochrocite, hydrated gibbsite, silica, and kaolin. The quantity of the hydrated sesquioxides decreases with depth until it merges into the pallid zone.

V The pallid zone. This zone appears to be most conspicuous in fine textured and/or less firmly cemented rocks. It is generally thin or indistinct in many profiles developed in coarse-grained sandstones and similar rocks. The zone is made conspicuous by its white to whitish-grey colour as well as by the intensity of the weathering process. The material is essentially clay, except in sandstones and other quartzitic rocks where the quartz grains are inert to normal chemical weathering processes. Finally, the pallid zone merges into the ordinarily weathered rock.

VI The ordinarily weathered or weathering country rock. In some cases the pallid zone may extend to the bedrock but this is not common. This last zone, unlike the others on top of it, retains the characteristic textures and structures of the country rock. The last zone is the unweathered country rock.

The total depth of the laterite (deep-weathering) profile varies widely. Measurements made in Nigeria are summarized in Table 8.4, p. 168. Table 8.3 shows the comparative depths of the indurated, mottled, and pallid zones for different geological environments. The weathered zone below the pallid zone is not included in the analyses but observation shows that this zone may be anything from a few centimetres to more than 40 metres thick. It represents, perhaps with the pallid zone above it, the C horizon, separated from the A horizon (I and II above) by a deep and complex B horizon (III and IV). The A and B horizons (Layers I–IV and possibly V) represent the solum, while the C and D horizons (V or VI–VIII) represent the parent material.

The formation of the laterite profile and the controlling factors as observed earlier, are not yet fully understood, being mostly inferred partly from the morphology of the profile and partly from conceptualization. A number of simple processes are thought to have combined to produce

TABLE 8.3
Deep Weathering in Nigeria: Profile Analysis

Sample location	Zone in profile	Mean thickness (m)	Standard deviation (n)	Coefficient of variation (%)
Jos Plateau, Rayfield	Indurated (I)	9·6	0·9	10
	Mottled (M)	19·9	3·6	20
	Pallid (P)	5·5	2·2	40
Jos Plateau, Barukin Ladi	I	12·3	38·6	310
	M	2·7	8·4	310
	P	3·7	9·4	310
Nigeria, Basement Complex country	I	1·9	3·2	280
	M	6·2	26·8	480
	P	7·2	26·0	360
Western Nigeria, sedimentary rock country	I	6·9	32·3	460
	M	27·1	120·4	460
	P	36·0	60·0	440
Mid-western Nigeria, sedimentary rock country	I	14·2	77·0	340
	M	39·0	202·6	520
	P	33·5	184·9	550
Nigeria, all sedimentary rock country	I	11·3	82·1	720
	M	34·4	243·7	720
	P	34·5	249·8	720
Total (all samples)	I	9·4	39·3	420
	M	21·3	100·9	470
	P	20·0	105·3	430

TABLE 8.4
Depth of Weathering in Nigeria: Some Statistical Measures

Sample location		Mean depth (m)	Standard deviation (n)	Coefficient of variation (%)	Chi-square	Degree of freedom	Probability	Significance
Jos Plateau (i), Dan Mongu area	Alluvium on granite, etc.	31·2	10·3	33	13·3	9	0·950	X*
Jos Plateau (ii)	Granite (thin alluvium)	10·3	7·9	76	12·7	7	0·950	X
South-western** Nigeria (i)	Sedimentary	66·9	19·8	29	1·3	6	0·975	X
South-western Nigeria (ii)	Basement Complex	13·9	3·0	22	0·3	4	0·999	X
South-western and northern Nigeria	Basement Complex	9·8	4·5	46	0·6	4	0·975	X

* X means not significant from normal distribution.
** The population studied consisted of a major and a minor one. The data here refer to the 'major' population only. The mean for all available data $= 86·9 \pm 35·6$; the coefficient of variation 40.

the profile as described. Some of them are unique while others are common to other soils. The unique processes are those that cause (i) the net loss of silica from the profile and (ii) the upward migration of large quantities of sesquioxides, especially aluminium and iron ions. Harrison's (1933) study shows conclusively the net loss of silica from the profile, while the occurrence of the zone of concretion (the indurated and mottled zones) on top of the pallid (sesquioxide-leached) zone suggests the operation of the latter process (i.e. capillary precipitation of sesquioxides at and/or near the soil surface). The two important conditions for silica leaching are high pH of the soil solution and the nature of the other cations and anions, but the detail of the process itself is yet to be understood. Part of the silica is lost to drainage water but a substantial part remains on land to silicify rocks within the vicinity of lateritization. The silcretes of Australia in particular were originally thought to be so connected.

The processes which are not unique to lateritization but which operate all the same are: the complete or near-complete leaching of the bases from the profile; the thorough decomposition of organic matter or humus; high degree of weathering resulting in essentially clayey soils lacking in primary minerals; and some (slight) eluviation of clay and sesquioxides from the A horizon. This horizon, especially where it is very thin, is readily removed after clearing and cultivation, except where conscious steps are taken to check soil erosion. Such steps are not usually taken with the result that the soils are destroyed quickly following the widespread practice of shifting cultivation and/or bush fallowing. Soil erosion and soil conservation are discussed in Chapter 10.

LATERITIC SOILS OF THE HUMID TROPICS
Included in this group of tropical soils are (a) soils with stratified profile morphology but no hardpan (duricrust or laterite) layer, and (b) soils developed on reweathered and/or truncated laterite profiles. The first category of soils includes those already described in the literature as ferralitic soils, ferrisols, latosols, etc. They consist of soils with a sesquioxide-enriched top horizon overlain by a humic, sometimes leached (eluviated) layer and underlain, as in laterite, by a mottled zone and/or a pallid zone. Also as in a laterite profile, the pallid or mottled zone is commonly separated from the fresh (unweathered) rock by a zone of ordinarily weathered or weathering country rock. The only difference is the absence of a duricrust or laterite layer, due perhaps to lack of desiccation. Such soils are therefore found in continuously wet high forest areas (Koppen's Af climate) of the humid tropical region. The humus zone is also generally thicker and may give the impression of a fertile black or brown earth.

The ferralitic profile contains in its upper ('indurated-zone') layer substantial quantities of both iron and aluminium hydroxides, and so is a type of laterite in the making. The compaction and induration of this layer will change it into a laterite profile. It therefore has many things in common with the laterite profile, for example:

(a) A low cation exchange capacity and a low base saturation in the colloids, owing to excessive hydration, rapid removal of bases, and low absorption and adsorption capacity of the clays.
(b) Greater amounts of sesquioxides in the upper than in the lower part of the profile.
(c) A smaller amount of combined silica throughout the profile than in the parent material.
(d) A predominance of the kaolin among the clays present, or a tendency towards this.
(e) Accumulation of silica-free iron and aluminium oxides in both crystal and amorphous forms.
(f) A range of colours varying from bright red to darkish brown due to the presence of iron oxides.

Also like laterite, ferrallitic soils are not the best for agriculture. The humus content is generally low and so it requires heavy applications of fertilizers to maintain fertility. However, because it has not got the hardcrust layer, root penetration is deeper. Tree plantations, among other things, therefore do much better on these than on laterite soils.

The second category of lateritic soils are those derived from reweathered laterites and related materials. Laterites in most places are very old, dating back to millions of years. They have therefore been widely truncated and/or destroyed. Examples of soils developed on the materials of the various zones of a laterite (duricrust) profile have already been discussed (Chapter 4). Observations from many localities show a toposequence of soils varying from the hilltops to the valley bottoms. Thus, while a typical laterite profile may occur on the hilltops and other interfluve regions where the original surface has not been destroyed, lateritic soils with no hardcrust horizon may occur along the valley-side slope of the dissected country. Finally in some valley bottoms, especially of ephemeral streams, deposited laterite debris may be reworked and formed into a type of laterite soil.

KRASNOZEMS AND RELATED SOILS OF THE HUMID TROPICS

The word krasnozem is of Russian origin. It is generally used for soils which appear to be red throughout, i.e. have undifferentiated soil profiles. The word was first used to describe such soils from the southern part of the USSR, and similar soils have since been reported in many parts of the humid tropics, particularly in Hawaii and humid tropical Australia.

Whereas in the extra-tropical areas krasnozems tend to develop only on iron-rich rocks such as basalt, those in the humid tropics occur on a wide variety of parent materials. The important factor, apart from climate and vegetation, appears to be good drainage. While laterites and lateritic soils occur in areas with impeded drainage of some sort, especially those on old erosion surfaces, krasnozems are associated essentially with well-drained, high-relief regions. Krasnozems may also develop on denuded and dissected laterite surfaces, in which case they are lateritic soils of some sort. Indeed, as observed earlier, laterites, lateritic soils, and krasnozems are only points on a continuum of continuously varying soil types.

The most important distinguishing characteristic of krasnozems is the uniformly red colour. It has therefore been referred to in many tropical regions as red earth. Profile differentiation is generally poor. The top part is 'coloured' by organic matter, making it brown. As in laterites and lateritic soils, humus decomposition is rapid and fairly thorough. There is therefore little or no O horizon, although roots may penetrate quite deeply into the profile. The humus/A horizon is generally light in texture, being mainly sandy loam or at the best clay loam. By contrast, the B horizon is mainly bright to dark red and generally more clayey. This horizon can be quite deep, especially in some sedimentary rock areas. Although illuviation may account for part of the clay in the horizon, weathering *in situ* is also important.

Krasnozems vary, particularly in depth, in different rock types. The deepest profiles in Nigeria were observed in the loose sandstone areas, especially the unconsolidated sandstones of the Cretaceous Age. The best and also the deepest examples were observed around Bida, Benin, Enugu-Nsukka, and parts of north-eastern Nigeria. In the Basalt areas of the Biu-Damaturu area of Northeastern State, the profile is generally much shallower, presumably because the soils are much younger than the others.

Sesquioxide enrichment is not as intense as in laterites and lateritic soils. Iron may make up 10–20 per cent of the soil, which proportion is quite low compared to laterites. This situation appears to be due to the fact that the iron is distributed throughout the profile, rather than in a particular part of the profile.

The process by which krasnozems form has been described as ferrallitization as opposed to lateritization. The process operates best in freely drained tropical areas which have both even rainfall distribution and distinct dry seasons (Koppen's Af, Am, and Aw climates). Thus, krasnozems, like laterites and lateritic soils, are common in the tropical rainforests, monsoon woodlands, and savanna woodlands. However, unlike laterites, a dry season and the presence of the water table within the pedogenic horizon are not essential. Indeed, these factors, if present, will lead to the formation of a laterite rather than a krasnozem or red-earth profile.

Another variant of the krasnozem is the yellow earth, in which the profile is uniformly yellow throughout. Yellow earth tends to occur mainly in the more siliceous parent materials containing the more hydrated iron oxides. Some sandstones and the pallid zones of truncated laterites and lateritic soils have also been observed to be particularly prone to yellow-earth formation.

Finally, and to illustrate the close connection between the different tropical soils described above, it is important to take note of two things. First, they can all occur in a catenary sequence, whether as active or as relict soils. As active soils, krasnozems tend to occupy the well-drained slopes while laterites form in poorly drained depression and plains, where a fluctuating water table occurs. By contrast, as relict soils, laterites occur on the hilltops, plateau-tops, and interfluves where the original profile is preserved intact. Truncation and stripping may expose the various materials of the laterite profile along valleyside slopes, leading to the formation of lateritic soils on weathered laterites and krasnozems at lower positions, particularly mottled-zone materials. Red earths tend to form on top of some mottled zones, especially the iron-enriched portion, and yellow earth on pallid-zone materials. Second, slight environmental changes can transform laterites and lateritic soils into krasnozems and vice versa. The iron in the zone of concretion (indurated zone) of a laterite profile may be redistributed throughout the profile following reweathering and renewed pedogenesis, while the occurrence of a groundwater and impeded drainage situation may lead to the formation of a typical laterite profile from a krasnozem profile.

The Soils of the Humid Temperate Regions

The humid temperate region, as the name implies, is an area where climate is not extreme; that is, it is never too hot, too cold, too dry, or too wet. The region was first defined as such by the Greeks who hypothesized that the other zones—the torrid and the frigid—are unsuitable for civilized habitation. Since the Greeks, eminent climatologists have tried to define and delimit this region. One of these was Koppen, who grouped the region under his C and D climates. However, for the purposes of this chapter, especially for the pedogenic process now being considered, the temperate lands will be limited to areas with cool climates, especially those characterized by forest vegetation, i.e. the area of the mid-latitude forests. This excludes the mid-latitude grasslands, to be described in the following section.

The humid temperate lands are significant for many reasons. Although occupying just about 7 per cent of the earth's land area, they contain over 40 per cent of the human population. The main concentrations are in the northern hemisphere, e.g. in China, Europe, and North America.

Moreover, the mid-latitude forest regions house the main centres of civilization at the present time, i.e. the oriental or Eastern (China) and the occidental or Western (European and North American) civilizations. The history of occupancy is also very old in some areas, e.g. in China and Europe, thus introducing the human factor into the process of soil formation. Indeed, perhaps the most important factor of the soils in this region, as also in the temperate grasslands (see below), is the extent of human interference. As a result of the human factor, the pedogenic process operating in this region is among the best understood. This is the process of podzolization, the essential features of which include: (a) severe base depletion throughout the profile, resulting among other things in the dominance of hydrogen clays, low pH, and marked profile (horizon) differentiation; (b) accumulation of a surface mat (O horizons) of acid organic matter, which considerably contributes to the intense leaching of the bases and the clays; (c) eluviation of iron and humus from the E horizon; and (d) illuviation of humus, iron, aluminium, and clays in the B horizon. Not all these occur in every podzol and podzolic profile, but at least one is necessary before a soil profile can be so described.

Originally, the word podzol, derived from Russian words meaning 'ashes underneath', was used to describe soils with an ash-grey A horizon, particularly the horizon immediately underneath the organic topsoil. This is the E horizon of the FAO/UNESCO system and is also called the *bleicherde*. Subsequently, however, the name podzol has been extended to include soils of widely varying profile characteristics.

Two main categories are commonly described: podzols and podzolic soils. The term 'podzol' is generally restricted to soils in which all the above-listed characteristics are present. By contrast, podzolic soils have at least one of the characteristics, especially the first one. Leaching of the bases is common in most humid soils, including the laterites and lateritic soils; however, this process is most pronounced in podzols and podzolic soils, particularly the former. Each of the two soil categories may be further subdivided. Variants of podzols include orthods, humods, humus-iron podzols, and aquods; those of podzolic soils include brown earths, red earths and yellow podzolics, and grey-brown podzolic soils. Each of these soil types will now be described.

Podzols
Podzols develop best where the climate is sufficiently cool to minimize the rate of litter decomposition, sufficiently wet to maintain an overall downward leaching of soluble soil constituents, and sufficiently warm to encourage enough humification to produce a highly acidic humus layer from which large quantities of fulvic acid are constantly released. This fulvic acid, being mobile, effects the complete leaching of the bases from the entire profile as well as the movement of iron and aluminium through

the soil profile. Illuviation of humus and iron often results in the formation of a structural B horizon. This may be thin and firmly cemented (ortstein) or thicker but less firmly cemented (orterde). A silicate clay cemented horizon or hardpan may also occur.

The climates associated with the areas of typical podzols have cold winters and short warm summers, i.e. Koppen's humid micro-thermal climates (Dfa, Dfb, Dfc, and to some extent Cfb). Other conditions which favour the formation of podzols are: (1) quartz-rich (silicic) rocks—sands and gravels, sandstones, glacial tills, siltstones, clays, and alluvial sands are the commonest parent materials for podzols—they are rarely found on basic rocks such as basalt, even under the most ideal climatic conditions; and (2) heath, coniferous, and boreal type vegetation. In Europe spruce, pine, larch, and birch are dominant; in Siberia the Dahurian larch; in the Far East fir and spruce and in North America spruce, fir, tamaract, etc. Light penetration to the forest floor is minimal, thus reducing the number and range of organisms that effect humification and mineralization. Earthworms are particularly rare; soil mixing, as it occurs in the humid warm regions, is therefore minimal, resulting in the formation of well-defined horizons.

The most important process operating in podzolization is the redistribution of the soil constituents by percolating groundwater. This water, charged with the fulvic and other acids from the O horizons, can dissolve or extract a number of soil constituents as it moves down the profile. These constituents include fine particles of organic matter converted into organomineral complexes by the processes of humification and mineralization, iron, and in some rare cases aluminium compounds, mineral clays, and silica. The organic matter is commonly illuviated immediately below the Ea horizon, in the Bh horizon. This is followed immediately by the deposition of iron and organic matter in the Bfe horizon. The Bfe horizon may be further subdivided, depending on the relative proportions of the iron and humus constituents. The Bs and B horizons which are mostly clay, enriched with iron, and are strongly acid, occur below the Bfe horizons in the most ideal cases. Finally there is the C horizon which is the soil parent material. The presence of groundwater may modify the profile by creating a grey Bs horizon (Bsg) or even a mottled B (gley) horizon, as in a groundwater podzol. In other words, variations in the effective factors as well as in the age of the profiles result in the existence of many types of podzol profile.

Of these factors, parent material is perhaps the most important. In a parent material rich in iron, a sequence of podzol profiles starting with an acid brown profile passing through an iron podzol to a humus-iron podzol with time has been observed; see Fig. 8.3 (a). However, parent materials lacking in iron cannot develop a humus-iron podzol profile. Such materials in ideal situations can develop a humus-podzol profile

WORLD SOILS 175

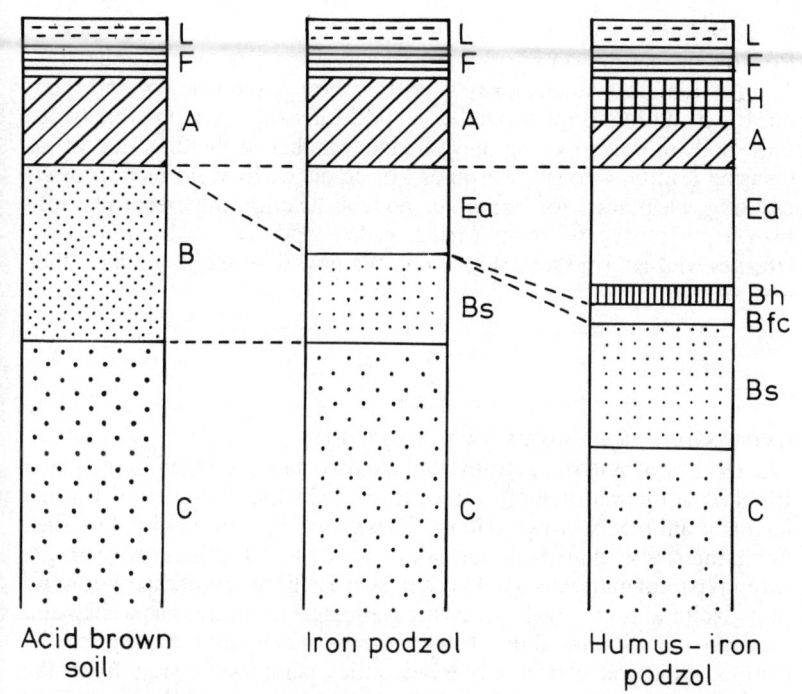

Fig. 8.3 Examples of podzol profile
(a) Maturity sequence in soils from acid brown soil to humus iron podzol

(humods), in which the iron pan (Bfe) and the silicified (Bs) layer may be absent. Another important factor is groundwater. Where a high groundwater occurs, a groundwater podzol or aquod develops. Here the B horizon may be enriched by iron and/or humus (organic matter). Alternate wetting and drying may cause mottling as well as gleying.

Podzols have also been observed to differ in profile characteristics in upland and lowland locations.

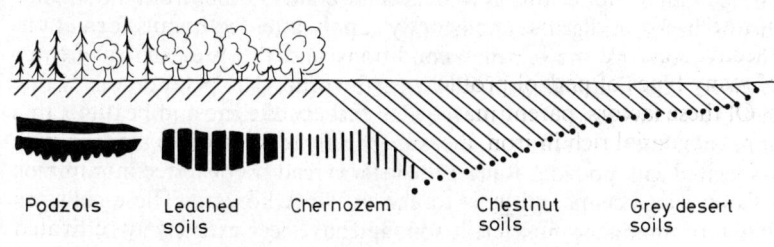

Fig. 8.3 (b) Climatic sequence of podzolics

Podzolic Soils

Podzols proper develop in the most ideal situations for podzolization. As the situation changes away from this core, a sequence of soil types occur which are grouped together as podzolic soils. It is possible to see these soils in succession as they change, whether in the direction of increasing aridity, wetness, warmness, or coldness. In the direction of increasing warmness, for example, podzols change imperceptibly into brown podzolics or brown earths, yellow and red earths, rendzinas, laterites, and lateritic soils (Fig. 8.3b). We now describe these and other soils.

BROWN EARTHS AND BROWN PODZOLICS

The term 'brown earth' is translated from *Braunerde*, which is the name Ramann, a German scientist, gave to the soils found in central Europe having a uniformly brown colour. These soils were originally described as characteristic of areas under oak or beech forest, especially over calcareous parent materials. However, brown earths now embrace a number of podzolic soils over widely varying geological formations as well as soil moisture conditions. Thus, brown earths of slightly varying profile characteristics have been observed under climates varying from the marine west coast type (Cfb) to the slightly continental micro-thermal climates (Dwb). Others are Dfa and Dfb climates. In these areas, winter temperatures average 0°C for about three months of the year, while summer temperatures range between 20°C and 26°C. However, perhaps the most important factor here is the rainfall, which occurs all the year round with a maximum during the autumn season. Drought is rare so that soil water is always sufficient to cause some leaching. But because temperatures and soil-water loss are higher than in the boreal-coniferous forest regions, soil water is usually not sufficient to cause podzolization *senso stricto*.

Because of the climate, plant life is more abundant and more varied. The leaf-fall of the deciduous trees and the contributions from the smaller shrubs, herbs, and grasses constantly supply litter for humus formation. This is aided by the warmer conditions and the presence of greater numbers of soil fauna, particularly earthworms. The plant debris of this region also has greater nutritive value than in conifers and heaths. Consequently, there is scarcely the superficial horizon of organic matter associated with podzols. Rather the litter is readily converted into humus as soon as it occurs and goes to enrich the A horizon. These soils are therefore among the most fertile soils and have been extensively cultivated all over the world. They rank with the mollisols of the mid-latitude grasslands in fertility and intensity of use, e.g. in the agricultural regions of Western Europe.

The following characteristics of the typical brown-earth soil profile explain the fertility:

1. Slight (minimal) leaching, particularly of carbonates, from the A and B horizons. The leaching may or may not affect the C horizon, where lime may be concentrated.
2. Neutral to moderately acid soil solution (pH 4·5–6·5). Both the base saturation and the pH increase with depth from the surface.
3. The humus is of the mull rather than the moder types and is present throughout the profile, particularly in the A horizon.
4. Clay eluviation is nil or slight; the clay fraction is therefore generally constant.

As with podzols and laterites, subtypes of brown earths are known, based on variations of parent material and on the intensity of leaching. The acid brown earths are commonly associated with sandy or silty parent materials, e.g. sandstones, siltstones, and coarse-grained granites; ferritic brown earths or brown podzolic soils occur on some reweathered ironstone (duricrust) concretions; while brown warp soils, some type of vertisols, are known to occur on alluvial deposits. The presence of a water table may cause gleying as well as varying intensity of leaching. Leached soils are commonly separated from brown earth proper, where leaching is minimal.

GREY-BROWN PODZOLIC SOILS

These soils are closer to podzols than the brown earths. The only differences are (1) there is no illuviation and precipitation of iron and so no Bfe horizons; and (2) features associated with organic O horizons and illuviation of humus are poorly developed. These they have in common with the brown earths. However, the more intense leaching activity and illuviation of clay, leading to the formation of an E horizon and a textural B horizon, distinguish them from brown earths and bring them closer to podzols. To distinguish them from both, they have been described as leached soils or alfisols (sols lessivé). The low humus and iron contents give the drab brownish-grey colour; as these constituents increase, the soil becomes more brownish until it becomes brown podzolic. Conversely, as leaching intensifies, it turns into a podzol. Grey-brown podzolic soils are therefore a sort of transitional (intermediate) soil type between the podzols and brown podzolics. Indeed, it is possible, as discussed earlier, to see all the soils of the humid temperate regions grading imperceptibly one into another, reflecting among other things the intensity of the pedogenic process of podzolization. The weakest and least developed profile is the *brown earth*. This grades through brown podzolic soils and the leached soils (grey-brown and red and yellow podzolic soils) into the podzols.

Grey-brown podzolic soils are widespread in the north-eastern USA, Central Europe, northern China, Manchuria, and Japan. They also occur within areas of podzols and red and yellow podzolics, especially where parent material is less suitable for any of the other soil types. Finally, in the transition from woodland to grasslands, grey-brown podzolic soils may grade into grey podzolic soils. Here, leaching is minimal and evidence of calcification begins to appear.

RED AND YELLOW PODZOLICS

As we move into warmer areas in the humid temperate regions, brown earths change into red and yellow podzolics. These soils therefore occur in the intermediate zone between environments conducive to the formation of podzols and brown podzolics, and those conducive to lateritization and/or ferrallitization. As such, iron accumulation is more than in podzols and brown earths but less than in laterites and lateritic soils. The accumulated iron in the B horizons gives the soil its name, since the red and yellow colours are restricted to the horizon rather than characteristic of the entire profile as in the red earths, brown earths, or krasnozems.

Like grey-brown podzolic soils, red and yellow podzolic soils are leached soils (sols lessivé); the main difference is in their geographical locations. While grey-brown podzolics tend to occur on the cooler and wetter side of the podzols, red and yellow podzolics occur on the warmer and wetter side of the podzols.

The red and yellow podzolic soil profile is characterized by both base depletion and marked eluviation of clay and iron. There is therefore a bleicherde (eluvial) horizon, which may be poorly developed. Because of the warmer conditions, the accumulation of organic matter in the O horizon is minimal. There is no eluviation of humus, which, as in the grey-brown podzolic soil, is accumulated in the A horizon. Warmer conditions also mean more intense rock weathering and the formation of clay *in situ*. Red and yellow podzolic soils therefore contain much more clay in their B and C horizons than other soils of the humid temperate regions. They are only surpassed in this connection by the humid tropical soils.

Red and yellow podzolics are found mainly in the Cfa and Csa (warm temperate and subtropical) climates. The colour of the B horizon is due mainly to the form in which the constituent iron is precipitated. Poor drainage may cause hydration of the iron and the formation of yellow podzolics, while red podzolics are found in well-drained areas. Besides, the nature of the soil parent material can be important. In the Sydney area, for example, where the present-day podzolic processes are operating on truncated deeply weathered and duricrusted profiles, yellow podzolic soils tend to form on uplifted pallid zones as well as on some laterites (yellow-brown podzolic soil). Yellow podzolic soils also develop on Hawkesbury sandstones which contain hydrated iron oxides. By contrast, the red Narrabeen and Wianamatta shales and also the mottled zones of laterite (duricrust) profiles tend to develop red podzolic soils. Finally, completely bleached pallid zones at valley bottoms tend to develop greyish-yellow to grey podzolic soils. Other soils associated with this region which are dependent mainly on parent material are the red Mediterranean or terra rossa, the red-brown earths, and the light yellow-brown soils or cinnamon soils.

The Soils of the Semi-Arid and Arid Regions

Unlike the case of the humid regions where precipitation is greater than evapotranspiration, thus facilitating leaching of soil constituents by downward-moving groundwater, the movement of soluble constituents in the semi-arid and arid regions is either negligible or predominantly upwards. This is due partly to the low intensity of weathering generally and the movement of soil solution towards the soil surface during evaporation and evapotranspiration. Many single processes operate—e.g. salinization, solonization, solodization, etc.—but they may all be combined and described under the process of calcification, used generally to describe the process which forms calcic horizons (usually within the B or C horizon) and/or a mollic epipedon at or near the soil surface. Mollic epipedon refers to a soft, dark, friable surface horizon containing a high percentage of organic colloids saturated with calcic ions. Depending upon the intensity of the process and the amount of available calcium ions, a hard crust (calcrete, caliche) may be formed from the mollic epipedon.

Generally, the calcic horizons tend to be thickest and deepest along the boundary between dry and humid climates (in the semi-arid regions), becoming gradually shallower, thinner, and more compact with increasing aridity. These soils were previously described as pedocals, because of the prevalence of lime, as opposed to the pedalfers of the humid regions. They are also being described as soils dominated by accumulation of silicate clays and/or bases, particularly the latter.

The term 'base' is here used for both the cations of the alkali metals (e.g. sodium and potassium) and alkaline earths (e.g. calcium and magnesium). The basic cations are very mobile and so tend to form soluble compounds (e.g. sodium chloride) more readily in the soil. This is why they are readily removed from the soils of humid areas, except where drainage is impeded, e.g. in depressions. They tend to accumulate within the profile of soils in dry areas. Thus the common factor of soils in this region is the accumulation of the bases somewhere within the soil profile.

The accumulation may occur in all horizons or in particular horizons, depending upon the peculiar and dominating process.

The accumulation of bases in the soil profile takes two main forms. They may be found as free salts (sodium chloride, calcium carbonate) in combination with soil anions, or as cations adsorbed to the soil colloids, e.g. silicate clays. Formation of free salts presupposes that the soil colloids are base dominated, i.e. the presence of free salts of a certain base element indicates a concentration in the soil of that element in excess of what is required to exchange the cations on the soil colloids. Free salts are recognizable in the field by their white colour, salty taste, and reaction to silver nitrate. By contrast, bases in the soil colloids are more difficult to detect, except by inference from high pH readings, e.g. pH 8·5 and over.

The bases that accumulate in the soils of dry regions originate from various sources. By far the most important source is perhaps the soil parent material. Feldspars contain sodium, calcium, and potassium in their structures; some alumino-silicates are linked with basic cations while some sedimentary rocks and preweathered materials contain free salts (e.g. calcareous shales) and/or other cations attached to the colloidal minerals. Invasion of coastal areas by saline water, either from sea spray, sea wash, groundwater, or irrigation water is another possible source of these bases. A third source is rainwater. Apart from the intrinsic salts it contains, rainwater collects basic substances from the air (blown by wind from the ground surface) as well as from roofs, leaves, and other surfaces where these may have been deposited.

The dry regions as conceived here comprise such wide areas as grasslands and deserts. Although the general tendency in these areas is for bases to accumulate in the profile, the amount, type, and position of accumulation in the soil profile vary from place to place, while the type and amount of available organic matter (vegetation or plant life especially) can be very decisive.

The most extreme form of the calcification process is obviously salinization. According to this process, salt accumulation dominates the other pedogenic processes and so obscures other profile characteristics. The soil so formed is called a solonchak. There are two forms of this, namely, primary and secondary solonchaks. In a primary solonchak, accumulation or salt accession within the profile is at the maximum and little or no evidence of any soil formation is evident. A secondary solonchak is caused by changes (natural or artificial) in the temperature/soil-water relationship. Such changes may cause a portion of the former soil profile to be evident, or otherwise cause some form of profile differentiation to occur. The profile so formed is thus somewhat polygenetic. Nevertheless both soil types are saline soils characterized by visible salt crusts; neutral pH if the salt is sodium chloride with pH 7, and negligible if there is any vegetation cover.

As the climate becomes more humid, leaching sets in and the amount of available vegetation increases. Before the effect of vegetation becomes apparent in the grassland and prairie regions, a series of desert soils—solonetz, solodized-solonetz, and solodic—occur. In solonetz, profile differentiation becomes more advanced than in solonchaks. The salts are redistributed through the profile and some of them are removed completely. Moreover, clay, humus, lime, and gypsum (where these are present) are eluviated, although this is usually not sufficient to produce an E horizon. The process which produces solonetz soils is called solonization. Stronger leaching removes the excess free salts from the profile and replaces some of the bases (e.g. sodium) by hydrogen, thus lowering the pH and resulting in the formation of an E horizon. This is the process of solodization, by which a solodized-solonetz soil, or in the most extreme cases a solodic soil, is formed.

Although these soils tend to occur in a continuum at widely separated locations in space, e.g. in the direction of increasing wetness, they can also occur in a catenary sequence along a valleyside slope. At such locations, solonchaks occur at the valley bottoms with impeded drainage, e.g. playa floors, followed successively at higher positions by solonetz, solodized-solonetz, and solodic soils.

The addition of organic matter as a significant factor in the process of calcification makes a lot of difference. This is the case in the grassland regions where the rapid growth of grasses and herbs and the generally warm and humid summer seasons facilitate a ready supply of nutritive organic matter. Moreover, the drought of late summer and the frosts of winter largely arrest the process of decomposition of the available organic matter. Consequently the loss of organic matter is minimized and humus formation maximized. Thus, depending upon the quantity and quality of the available bases and organic matter as well as on the effectiveness of the humification and leaching processes, different types of soils may be formed. These soils are here described under two main subgroups, namely, the soils of the interior grassland regions and those of the desert regions.

Soils of the Interior Grassland Regions

The soils that come under this category are prairie soils, chernozems and other black earths, chestnut soils, and sierozems. Prairie soils are found essentially where the native vegetation is tall grassland in the transitional zone between dry woodlands and real savannas or grasslands. In North and South America, these soils occur between the podzolized and lateritic (forest) soils to the east and the chernozems and black earths (grassland soils) to the west. The soils resemble the chernozems except that lime is removed from their solum. They are therefore transitional between the pedalfers of the wet regions and the pedocals of the dry regions. Moreover, prairie soil peds can be quite hard when dry because of the favourable conditions for both flocculation and cementation. They also contain high iron contents and so are sometimes grouped with the pedalfers. The iron is frequently coloured by humic acid which gives the black pigmentation, unlike the red and yellow colours of most pedalfers. However, like most pedocals, the profile is not usually very deep, reflecting their moderate weathering environments. The profile also has two distinct horizons as in chernozems—the Ah and A (or A and B) horizons overlying the C horizon. They are also best developed in areas of free drainage. Leaching of the bases and humus is not intense, although more than in chernozems. Prairie soils, like chernozems, are very productive. Apart from having most of the desirable features of black earths and chernozems, they also occur in much wetter regions, thus minimizing the limiting effect of this important factor of cultivation.

Chernozems and black earths are used synonymously in most elementary textbooks but pedologists tend to distinguish between them. This is usually done in relation to either the organic matter or the clay content of the upper horizons or solum. Chernozems are commonly defined as having more organic matter and less clay content in the upper horizon than black earths. The dividing marks are 4 per cent for organic matter and 50 per cent for clay content.

The formative processes of both soils are, however, similar. The processes are intimately related to the grassland vegetation of the region. This vegetation plays an important role both in the formation of humic acids and in the return of calcium (bases) to the soil surface. Humus eluviation is negligible, if any, as humus is always plentiful at the surface, decreasing with depth. Clay eluviation is also unimportant and clay may be quite abundant near the surface. Free alkali salts as well as alkali cations (sodium and potassium) are generally removed from the profile, although some alkali cations may remain in the colloids in the lower B horizons.

Chernozems and black earths are among the most fertile and definitely the most extensively cultivated soils in the world, particularly in Central Europe, the North America prairies, Argentina, the Murray Darling area of Australia, and South Africa. The areas occupied by these soils in the temperate regions are popularly known as the granaries of the world, just as the tropical grasslands (savannas) are important in the production of grain crops. While wheat, barley, and oats feature in the temperate regions, millet, sorghum, and guinea corn feature in the tropical lands, especially in tropical Africa.

Chestnut soils are the soils of the drier parts of the short-grass areas or steppe (BS) climates. Rainfall and vegetation are much less than in the chernozem region. Horizonation is generally indistinct; and the surface horizon is grey-brown, friable, and platy structured. The calcic horizon is nearer the surface than in chernozem, while the slower decay rate of a smaller supply of plant debris produces less humus. Chestnut soils are extensive in the extreme south of the Ukraine and in a broad arc from the western shore of the Caspian Sea eastwards along latitude 50°N, to the Irtysh River in Central Asia. In North America, these soils extend from the North Saskatchewan River south-eastwards to the Llano Estacado on the high plains east of the Rockies. Crop cultivation is not as important as in the chernozem and black-earth region, but grazing is very important. Chestnut soils have been described as grey and brown soils, especially as they grade into the grey desert soils or sierozems. These soils have their upper horizons enriched by calcium carbonate. Vegetation is scarce, mainly sage bush and bunch grasses which contribute very little humus to the soils.

Soils of the Desert Regions

True deserts (BW climates) cover a large part of the earth's land surface. The most extensive areas are those of Africa (Sahara and Kalahari), Eurasia, and Australia. These are regions with very extreme climates, so extreme in some places that little or nothing happens in terms of soil formation. Weathering is mainly of the mechanical type which produces essentially coarse regolith. Some chemical weathering occurs and, with the sorting action of winds and moving water, the soil parent material is well differentiated into rocky, gravelly, and sandy areas. Playa floors have silt and clay-size particles in addition.

The most important process of soil formation is the vertical movement of the salts in the soil profile. We have listed the individual processes as salinization, solonization, and solodization, which operate at suitable locations, resulting in the formation of solonchaks, solonetz, and solodic soils respectively.

Solonchaks are also called white alkali soils mainly because of the colour of the surface horizon, due to excess sodium salts. The most suitable environments for their maximum development are low-rainfall coastal areas (especially those which are inundated by saline water periodically) and the playas. In such places conditions are ideal for salt accession, which far outweighs salt removal by the leaching processes. The result is a complete (in primary solonchak) or almost complete (in secondary solonchak) masking of the soil profile by salt accumulation. At best a profile consisting of a surface crust of salt underlain by a sodium-saturated clay horizon is formed. Such soils are definitely useless for any cultivation, not only because they are dry but also because of the excessive salt (sodium) content.

Horizonation improves with increasing wetness, due to increased rainfall or to other sources. For instance, in solonetz soils the salt is completely dispersed throughout the profile rather than flocculated as in solonchaks. Solonetz soils are, however, just as poor as, or even poorer than, solonchaks, on account of their peculiar physical (morphological) characteristics. Where there is sufficient vegetation to provide humus, staining may occur through humus being eluviated. In America such soils are called black alkali soils. Because plant is available also, grazing is practised but reclamation (e.g. by irrigation and addition of gypsum) is essential for any worthwhile use. Finally, better drainage and increased wetness intensify the pedogenic process to form either the solodized-solonetz or the solodic soils, neither of which is useful for cultivation without reclamation of some sort.

The conditions which give rise to solonchaks and related soils are not widespread in desert environments, as playas and periodically inundated coastlands form only a small part of desert areas. Indeed, such environments are not real deserts, at least not as dry as most interior locations.

In these places soils form in either sands or loamy materials. It is therefore possible to recognize two soil categories—the desert sands and the desert loams. In both these soils, the individual soil particles are commonly coated by oxidized iron, which gives the soils a red colour. Profile development is negligible and the soils tend to be structureless, except that the high permeability of sands facilitates greater base removal. This lowers the pH of such soils.

The Soils of the Cold Lands

The cold lands are found in two main environments—the high-altitude and the high-latitude areas. These are the areas of the subarctic (Dwc, Dwd, Dfc), ice-capped (ET), and tundra (ET) climates of Koppen. Accordingly, there are two distinct pedological environments demarcated by the tree-line. The ice-capped and cold desert regions have no soil as such.

The tundra (ET) and the subarctic occur to the largest extent in the northern hemisphere, surrounding the Arctic Ocean, but are restricted mainly to high-altitude areas in the southern hemisphere. They are characterized by a severe climate with long, bitterly cold winters and short cool summers.

The most important controls on soil formation are climate and vegetation. Soil formation occurs mainly during the brief summer when the soil surface thaws. Below this thawed material is, of course, the permafrost, or permanently frozen ground. Furthermore, the thawed soil is usually saturated with water, owing to the impedance created by the underlying permafrost. Lateral movement occurs downslope (solifluction) and may involve some redistribution or sorting of the soil according to particle size. The vegetation consists essentially of heather, arctic blueberry, buttercups, mountain avens, lichens, and mosses. In the subarctic there are scattered stands of stunted conifers. The rate of plant growth, and so of organic matter supply, is slow, but the rate of humification is even slower, so that there is an accumulation of raw organic matter on the soils. In other words, the soils of the cold lands can be distinguished on the basis of these factors of drainage and humus content. These are usually affected by micro-relief.

In the first place the depth of the effective soil is controlled by the position of the permafrost, relative to the surface. Where it is close to the surface, downward leaching is negligible, if at all possible. Horizonation is therefore ruled out or is considerably minimized; the result is bog soil. As the thawed zone deepens, profile development is facilitated, but this is only evident on the outside margin of the tundra, i.e. the subarctic region as it merges into the boreal forest region. The succession of soil types from the most drained to the least drained, i.e. based on the degree of hydromorphism, are eluviated tundra soils, meadow tundra soils, tundra gley soils, half bog soils, and bog soils. By far the most widespread of these are the tundra gleys, developed on gently undulating plains and other similar surfaces. They are comparatively deep soils, with a peaty upper horizon overlying a mixed layer of mineral matter saturated with water. The colour of the A horizon is typically yellow-brown to brown, reflecting the influence of the organic matter. That of meadow tundra soils is dark grey, reflecting the poor drainage of the low-lying locations. As the depth of the effect soil or the permafrost decreases, drainage worsens and the soils turn into tundra gley, semi-bogs, and finally bogs.

Horizonation may, however, be impeded by other factors. Cryoturbation is one such process. This is the churning activity which occurs within the active layer above the permafrost. This occurs particularly when a water-saturated layer occurs between the permafrost and a frozen surface layer, during an ensuing winter. The expansion caused by freezing builds up pressures which may rupture the surface and cause the underlying material to be spilled to the surface. Solifluction, by keeping the active layer in motion, also inhibits development of horizons in many tundra soils. However, solifluction effects sorting and so can produce a soil catena or topo-sequence of some sort. In such a situation, upper slopes are characterized by rugged, frost-shattered rock with little or no soil. Screes occur immediately below this with the particles becoming finer with distance from the hilltop. The finest particles occur at the footslopes and alluvial material occurs in valley bottoms.

Finally, there is the factor of organic matter. Depending upon the nature and quantity present in soils, tundra soils have been classified into types. The most commonly recognized types are alpine humus soils, acid moor soils, peaty gley soils, and peats.

The best-drained soil is the alpine humus soil. As in most tundra soils, the B horizon is usually absent while the A horizon is divisible into A1h, A2h, and AC. The top is littered by plant debris which supplies the humus to the A horizons, the amount of which decreases with depth. The dominating pedogenic processes in alpine humus soils therefore are base depletion, the accumulation of a weakly acid mull litter, and the incorporation of humus from the litter into the A horizon. Since frost shattering is the dominant weathering process, the soils are generally gravelly.

As the water table rises, saturation increases and degree of leaching decreases. Should the temperature decrease at the same time, as it does on the poleward side of alpine humus soils, rate of decomposition decreases, and the organic matter content of the soil increases. Where there is sufficient relief, leaching intensity is still high but not as in alpine humus soils. Acid moor peats replace alpine humus soils in such environments, if the groundwater, and so soil solution, is acidic; or alkaline peats, if the groundwater is alkaline. Finally, if humus accumulation overlies strongly mottled subsoils, the soil is called a peaty gley soil or humic gley soil.

There is no soil worth describing in the ice-cap (EF) climates and so we shall not spend any time on that area. Even in the case of the tundra and subarctic region, the soil is not suitable for cultivation except under special management conditions. Apart from the permafrost setting the limit for root penetration, the soils are almost always saturated with water and in constant movement. Lack of chemical weathering means little or no clays but mainly coarse pebbles. Organic matter is also negligible in many places, except perhaps in the alpine humus soils. Thus, while it is possible to turn certain deserts into rich farming land by irrigation, reclamation of tundra soils is much more difficult.

Relict Soils

The basic assumption in the soils described so far is that they developed under a constant environment, particularly the present-day climatic and biotic environments. This is also the basic assumption of most authors in this field, many of whom failed to recognize the significance of marked fluctuations in climates, particularly those of the immediate past. Among other things, climatic change implies changes in vegetation, weathering, and pedogenic regimes, as well as erosion and depositional characteristics.

The implications of changes in the pedogenic environment are many, and only a few can be mentioned here. For example, it is possible to imagine the following situations in connection with relict soils:

(1) The occurrence of relict soils formed many millions of years ago, e.g. duricrusts, especially those referable to the Tertiary period.
(2) The superimposition of present-day pedogenesis on previous soil profiles, e.g. the podzolized laterites of vast areas of the mid-latitudes.

(3) The occurrence of buried soils, especially in depositional and volcanic environments, leading to the formation of complex or polygenetic soil profiles.
(4) The occurrence of arid soils (of the recent arid period) in areas which are today much wetter.

We shall now discuss each of these situations very briefly.

Relict Duricrust or Deep-Weathering Profiles
The argument as to whether deep weathering and duricrusting is a geological or a pedogenic process has long since died down; indeed the typical deep-weathering profile has already been shown to be similar to a humid tropical soil profile—e.g. the laterite profile. Also the ferruginous tropical soil of the French school has the same profile characteristics as a deep-weathering (duricrust) profile. This profile or phenomenon is characteristic of a particular weathering environment. Thus where the phenomenon occurs outside the humid tropics of today, it should be a relict of a humid tropical environment.

Perhaps the most important area for this phenomenon, because of the extent of documentation, is Australia, where it has been assigned to the Tertiary Age. Except in parts of the Northern Territory and possibly also northern Queensland, the climatic environment over large parts of Australia at the present time is too dry and too cool for deep weathering and duricrusting. Yet this phenomenon occurs widely, even in Tasmania. Two types are known: the ferruginous type in the wet north, east, and south-west, and the siliceous type over large parts of the remaining (dry) area.

It is difficult to say precisely what the environment was in Australia when these materials were formed. Present-day conditions may be poor guides, especially given the epeirogenic episodes (Kosciusko uplifts) which the eastern parts of Australia witnessed during the Pliocene–Pleistocene period. Moreover, the actual conditions which promote mobilization of the sesquioxides of iron and aluminium on the one hand and of silica on the other are not clearly known. The pH of the soil solution is important; so also is the Eh and the organic matter content. One thing that is certain is that the duricrusts in the wet coastal areas are mainly ferruginous and/or aluminous (e.g. ferricrete, laterite, and bauxite) while those in the drier inland locations are essentially siliceous (silcrete or silcrust)(Fig. 8.4). The suppositions either that the silcrust could be the pallid zone of a ferruginous/aluminous duricrust profile or that the silica was transported by water from the sesquioxide-enriched areas are now largely discredited, following the discovery of convincing evidence that the silcrusts are similar to the laterites, etc. in many respects, including their physical and morphological (profile) characteristics. Both materials may

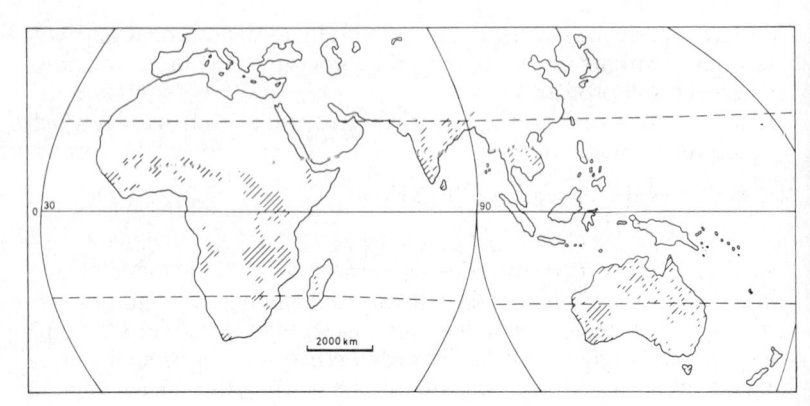

Fig. 8.4 The major occurrences of duricrust

be pisolitic, vermicular, vesicular, or massive, while the three zones—indurated, mottled, and pallid—are equally present. The main differences are in the depth of the profiles, the silcrete profiles being much deeper than the ferruginous-crust profiles, and in terms of mineralogy. Moreover, the profiles in the wetter and also less stable areas along the east coast have been more widely truncated and/or destroyed than those in the dry and more stable inland regions. Present-day pedogenic processes have also been more active in the wet areas than in the dry areas. Thus while the crusts and related materials have been reweathered and podzolized in wide areas as described above under lateritic soils, these materials are still mostly intact and fresh in the dry (interior) areas.

Deep-weathering and duricrust profiles are also widespread in other continents. In North America, the bauxites and related materials of Arkansas and other areas in the south-eastern USA are typical duricrusts with characteristic deep-weathering profiles (Harder, 1952; Sherman, 1952; Gordon et al., 1958). Furthermore, typical deep-weathering profiles have been described in north-west Oregon and Wisconsin (Allen, 1948; Dury and Fox, 1971). Dury and Fox wrote of the Wisconsin situation as follows (p. 291):

> Duricrusts and deep weathering profiles are exposed in numerous sections in Southwestern Wisconsin.... They transgress the local sedimentary succession, from not lower than the St Peter sandstone ... to well down into the Cambrian. The profiles include pallid zones in their lower parts, with mottled zones next above, and duricrusts at the top, being wholly comparable in their characteristics and relationships to the profiles and crust abundantly described from the Southern Hemisphere, notably Australia.

Work on the North American examples is still in progress; so far they have been ascribed with the others in the other continents, to not later than the Tertiary period. Similar dates have been suggested for the deep-weathering profiles in north-east Ireland, western Germany, etc., and the bauxites and similar materials of southern France, among other places in Europe (Cole, 1908; Cole et al., 1912; Fox, 1932; Dury, 1971).

In Africa, duricrust, mainly of the ferruginous and/or aluminous types, caps extensive areas of plateau-tops, hilltops, and plains. These materials have been widely used in stratigraphical studies, e.g. to recognize erosion surfaces (King, 1950). Although deep weathering and duricrusting are still feasible in some areas of tropical Africa, all available evidence suggests that the duricrusts, particularly the primary and some of the secondary ones, are fossil, referable mainly to the Tertiary and earlier periods.

Wherever these soils occur, they have similar characteristics, including the occurrence, essentially at the surface, of a hard crust, hardly penetrable by roots. Also, being the result of deep tropical weathering, the soils are depleted of all bases and, except where vegetation provides a ready supply of organic matter, humus. Indeed, duricrusted surfaces especially in dry environments are virtually barren areas, hardly used for any cultivation. Where such areas have been dissected and the underlying softer (mottled zone, pallid zone, and ordinarily weathered rock) materials are exposed, soils which vary in catenary sequence as described earlier have been formed, which support extensive plant life and sometimes crops. This happens in most wet areas where even the duricrust has been reweathered, transported, and redeposited to form the parent material of subsequent soils. In such situations, the new soil profiles are 'superimposed' on the former soil profiles. In some cases the subsequent processes are strong enough to change completely the character of the original materials; in others the change is only partial. This topic has already been discussed earlier in this chapter.

Buried soils are other examples of relict soils. In this case, changes in the environmental setting, instead of resulting in the dissection and truncation (erosion) of existing soil profiles, cause such profiles to be buried. Burial may be by alluvial, colluvial, aeolian, glacial, beach, or lake deposits or it may be the result of lava (basaltic) flow. Like duricrust, buried soils have been widely studied in Australia, where the K-cycle theory actually emanated (Butler, 1959). The soils of the Riverine Plain—the region traversed by the Murray, Murrumbidgee, Goulburn, Campaspe, and Loddon rivers and their tributaries—are among the best described of such soils (Butler, 1958). Here a complex relationship exists between wind- and river-deposited load on the one hand and the deposits by the former and the present-day streams on the other. Furthermore the buried soils have themselves been modified, either by truncation or by subsequent pedogenic activities.

190   ESSENTIALS OF SOIL STUDY

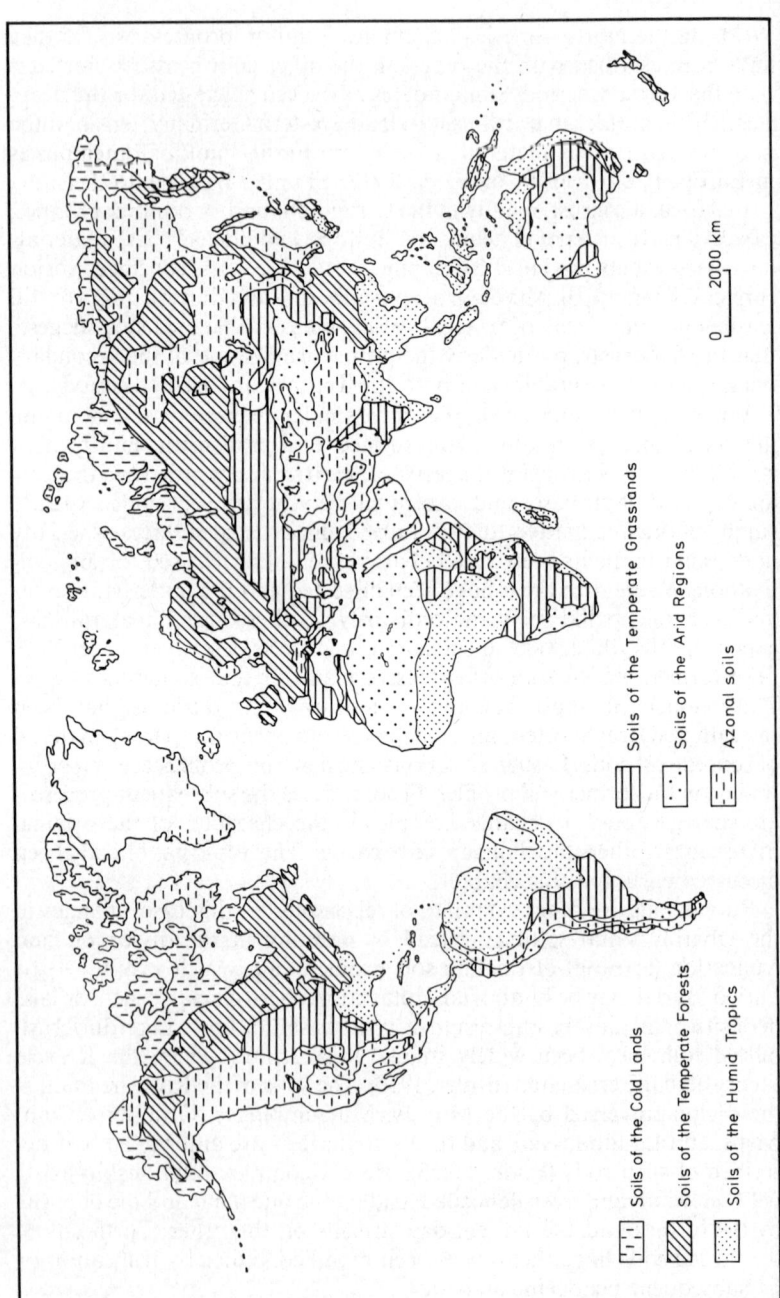

The best examples of buried soils known to the authors were observed in Nigeria. Here laterites and lateritic soils have been buried under thick alluvial, eolian, lake, and basaltic deposits over wide areas. Examples of such buried soils are particularly widespread in the Chad basin where they are covered by the Chad formations and in the Biu Plateau areas of north-east Nigeria where they are buried under basaltic flows. Buried soils are of little use except when they are exhumed or otherwise worked into the present-day soils.

Finally, there are areas where the soils of the recent arid period are preserved. These soils or materials have been widely used in the Quaternary chronological study, particularly in West Africa. Among the important works in this respect is that by Burke and Durotoye (1970), based on Bruckner's (1955, 1957) pioneer work in Ghana. However the effects of previous aridity are generally much less preserved in soils than the effects of previous greater rainfall. While desert environments preserve whatever is there for posterity the materials (salts, etc.) accumulated during aridity are readily leached on the advance of humid conditions. If the change is not marked, the effects of the arid process may not be completely obliterated, as in solodic-solonetz and solodic soils, or in degraded chernozems, i.e. chernozems affected by mild podzolization.

Regosols, Lithosols, and Raw Mineral Soils

There are wide areas of the earth's surface where soil formation either has not started or has not been effective. This may be because the processes have not operated for a long enough period or because other factors, e.g. steep slope, have not allowed the soil to accumulate. Examples of such soils are shown on the CCTA soil map of Africa, referred to earlier. On that map two types of raw mineral soils are shown—the rock and rock debris and the detritus, particularly the desert detritus. River-deposited materials or alluvium are commonly classified as regosol, presumably because it is assumed that they show some evidence of soil formation, being in a humid environment.

Large areas with no soil development or with weakly developed soils include beaches with fresh sand and shingle deposits, deserts (hot and cold), river valleys, alluvial fans, high mountain areas (particularly those in hard fresh rock), ice-cap regions and recent lava flows. Some of these areas are shown in Fig. 8.5, where the world soils are shown in a generalized and highly simplified form.

Chapter 9

SOIL-PLANT RELATIONS

Introduction
So far we have been looking at soil as a natural body, the way pedologists or pure soil scientists look at soils. However, the knowledge of soils for its own sake is useless. Indeed perhaps the obvious and, for practical purposes, the most important conception of soil remains that of soils as a medium for plant growth. Thus Buckman and Brady (1966) define soils as 'the natural bodies on which plants grow'. This is the edaphological approach to soil study, which emphasizes the potential or the ability of the soil to support plants generally and crops in particular. The term used in this connection is 'soil fertility', the major concern of this chapter. In the last chapter the implications of soil use are considered under the title 'Soil Erosion and Conservation'.

Factors of Soil Fertility
There are six environmental factors which influence plant growth: light, mechanical support, heat, air, water, and chemical elements. The soil is the medium through which plants obtain these requirements with the exception of light. The four factors, mechanical support, heat, air, and water, constitute the physical basis of soil fertility. They are related mainly to the physical properties of the soil such as texture and structure. In order to focus attention on the importance of structure in particular, the concept of soil tilth has been formulated. The chemical elements constitute the chemical basis of soil fertility. There are two major groups: first, there are the organic nutrients, nitrogen, phosphorus, and sulphur, associated with soil organic matter; secondly there are the metallic cations, calcium, magnesium, and potassium, derived from inorganic sources.

Mechanical Support
The soil is the rooting zone of plants. Plant roots serve in absorbing water and nutrient elements from the soil into the plant; in storing food and in fixing or anchoring the plant to the ground. The factors which influence root development in the soil include pore spaces and pore channels, temperature, air, and water supply, the amount of food (carbohydrates) distributed to the root system by the plant and competition between different plant root systems.

Roots grow and branch profusely in a soil with enough channels for root penetration. In hard, compact soils such as iron pans, or in heavy, fine-textured soils like clay, root penetration is largely through cracks and fissures. This is either because there are very few pore spaces (as in the pans) or because the pores are not wide enough for roots to penetrate (as in the clays). In soils with pans, plant roots tend to be concentrated in the soil layer above the pan and they usually grow and spread sideways rather than downwards.

The optimum temperature for plant root growth has not been studied in detail; it varies from plant to plant and from place to place. It is known, however, that plant roots thrive where the soil is moderately warm, well above freezing point in the temperate zones. Temperature affects plant roots indirectly through its influence on chemical and biological activity in the soil and in the loss of water from the soil through evapotranspiration. Evaporation from a warm soil will normally be at a higher rate than that from a soil with a lower temperature.

Good soil aeration is essential to plant root growth. Buckman and Brady (1966, p. 242) have defined the condition of good soil aeration as one in which gases are available to growing organisms in sufficient quantities and in the proper proportions to encourage optimum rates of the essential metabolic processes of these organisms. Soil aeration is dependent on the size-range and quantity of soil macro-pore spaces through which gases can move and diffuse freely within the soil and by which soil air can be constantly replenished from the atmosphere. Good soil drainage which would leave the macro-pores free of water is also essential. Good aeration facilitates the absorption of water and nutrients by plant roots from the soil and is essential for the functioning of soil microorganisms and for the decomposition of organic matter.

The composition of the air is a very important factor. Plant roots take up oxygen from the soil air and deposit carbon dioxide and other injurious gases. They are thought to require at least 5 per cent oxygen concentration to grow at all and between 9 and 12 per cent to avoid adverse effects (Russell, 1963). All plant roots are destroyed by high carbon dioxide concentrations. They grow best when the carbon dioxide level is less than 1 per cent, but considering the fact that the carbon dioxide content of the soil air fluctuates very much, plant roots must be able to withstand short periods of high concentration. The sensitivity of plant roots to carbon dioxide concentration tends to increase with temperature: at high temperatures they require greater quantities of oxygen to cope with the increase in the metabolic rate and so they require less concentration of carbon dioxide. In general soil aeration deteriorates with depth. The reasons for this may include the reduction in total pore space and in the average size of pores coupled with the great distance from the atmosphere so that there is less gaseous interchange.

Water supply is also essential to root growth. Plant roots commonly grow in search of water just as the aerial parts of plants grow towards the source of light. In dry soils found in arid areas, plant roots reach down to great depths in search of water. Indeed all plants, whether in arid or humid areas, play this vital role of bringing water and nutrients from the subsoil to the topsoil.

Heat and Air

The gases which occupy the macro-pores in the soil form the soil atmosphere, from which plant roots and soil micro-organisms obtain oxygen for their metabolism and deposit carbon dioxide and other obnoxious gases. Three important aspects of soil air are important in soil fertility: (1) the composition of the soil air; (2) the soil air/moisture ratio; and (3) the ease with which air circulates in the soil (diffusion).

The relative concentrations of oxygen and carbon dioxide in the soil depend on the rate at which plants and other living organisms take up oxygen and deposit carbon dioxide, the rate at which soil air is being replenished from the atmosphere (gaseous exchange), and the soil moisture condition. The rate of carbon dioxide production under natural conditions is not very well known but in the temperate latitudes it is estimated to be up to 10–12 grammes or 5–10 litres per square metre per day (Russell, op. cit.). This rate varies with the seasons and the type of crop. The presence in the soil of decaying organic matter or freshly applied green manure can increase the carbon dioxide content considerably: the micro-organisms become very active, thereby producing large quantities of carbon dioxide. This is a case for not planting crops too soon after the application of green manure: the high carbon dioxide content which reduces the relative proportions of oxygen and nitrogen may inhibit seed germination and damage young plants. Even on land carrying fully-grown crops, the initial effect of the application of green manure is the lowering of soil pH.

The carbon dioxide content of the soil is disposed of in two ways. Some of it is removed from the topsoil and rooting zone of plants into the deeper layers of the soil. Most of it is, however, sent to the atmosphere during gaseous interchange. The rate of exchange of gases between the soil and the atmosphere is influenced by the porosity of the soil. It is facilitated by temperature and pressure differences between the soil air and the atmosphere which initiate a mass flow of gases from one medium to another.

The total pore space in the soil is shared between gases and water molecules (Chapter 2). It follows therefore that the water relations of the soil will determine the amount of the air space. If the soil water content is excessively high this will have repercussions not only on the amount of air in the soil but also on its composition. In wet conditions, the carbon dioxide content of the soil increases, although some of it is in solution in the soil water. By contrast, the oxygen content decreases and none can be found in solution in the soil water. This is because any dissolved oxygen is quickly used up by micro-organisms leading to further activity and higher carbon dioxide concentration.

Partial pressure[1] differences within the soil determine, to a large extent, the movement and diffusion of gases into and within the soil. Even if there is no general pressure gradient, differences in the concentrations of one gas, say, oxygen, between the soil air and the atmosphere or between one part of the soil and another will result in the mass movement of that gas, usually from the area of higher to that of lower concentration. Diffusion is also directly related to the total air space available in the soil. This explains why the oxygen content tends to decrease and the carbon dioxide content tends to increase with soil depth. Topsoils are usually coarser in texture than the subsoil and the total air space and average size of pores tend to be greater. The worst affected are the subsoils of heavy, fine-textured soils in humid areas. Animal burrows and cracks in the soil greatly facilitate diffusion. A number of steps can also be taken to improve diffusion and thereby the composition of the soil air especially in the deeper horizons. These include (1) draining the soil to increase air space relative to water space, (2) deep ploughing and subsoiling to open up heavy soils, (3) maintaining a stable soil structure through organic matter by adding farmyard manure, and (4) growing leguminous plants.

Poor aeration retards the growth of plants and plant roots and impairs their ability to absorb water and nutrients. In the anaerobic conditions that prevail the acidity of the soil increases; certain mineral elements, e.g. Fe, Al, Mn, become very soluble and swamp the plant roots, thereby interfering with or preventing the uptake of more essential elements. Certain inorganic compounds are also produced which are toxic to plants. The micro-organisms in the soil are less active. Symbiotic nitrogen-fixing bacteria which depend on the oxidation of mineral and organic elements for their livelihood cannot survive. Only the anaerobic micro-organisms can function properly, extracting oxygen from mineral elements and compounds in the soil and in the soil solution. The rate of decomposition of litter is also slowed down and the products of such decomposition under anaerobic conditions differ from those produced under conditions of good aeration. They are mostly organic acids. Nitrogen compounds in the soil exist in the reduced forms which are not utilizable by plants.

Soil temperature is closely linked with soil air and it controls many physical, biological, and chemical processes in the soil (see Chapters 2 and 4). It is to a large extent determined by atmospheric temperature. In

---

[1] The partial pressure of a gas is the pressure that gas would exert if it were the only constituent of the volume of gases.

porous soils which allow free movement of air and gaseous exchange with the atmosphere, the temperature is in equilibrium with that of the atmosphere. In heavier, less well aerated soils, this equilibrium is likely to be between the atmosphere and the top layer of the soil only.

The temperature of the soil at any given time depends on the difference between the amount of solar energy absorbed and the amount of energy that is lost by radiation, conduction, and as latent heat of evaporation. The amount of energy absorbed by the soil in any given site is determined by such factors as (1) the amount and angle of solar radiation incident on the ground surface, (2) the slope angle and aspect, (3) the vegetative cover, and (4) the colour of the soil surface. The moisture content of the soil is also important: wet soils take longer to warm up than dry ones, because the specific heat of water is about five times that of inorganic soil particles. This explains the relative coolness of poorly drained soils especially in the deeper, less well aerated horizons.

Temperature is important in physical weathering, e.g. frost and freeze-thaw action in cold areas and rock expansion and contraction in hot areas with a high diurnal range of temperature. Temperature is also a major factor in evapotranspiration. It influences in one way or another such processes as photosynthesis, osmoregulation by plant roots and the uptake of nutrients, and the viscosity and movement of the soil solution—all of which are important to the functioning of plants. The rate at which micro-organisms decompose organic materials or fix atmospheric nitrogen is also dependent on an adequate supply of heat (temperature).

Water

Water accounts for a large proportion of the living weight of plants. For example, in many herbaceous species water forms 70–90 per cent of the fresh organic weight; and 40–60 per cent in many woody species (Watts, 1971). The singular characteristic of soil water which sets it apart from other materials absorbed by plants is that it is not returned in large proportions by the plant but flows continuously from the soil, into the root hair, up the stems into the leaves, and out on to the leaf surface from where it is evaporated. It is therefore the medium by which plant nutrients are carried in solution to all parts of the plant as well as the means by which heat is distributed and regulated in the plant. The ionization of elements in solution greatly facilitates chemical reactions within the plant. Water is important for sustaining plant protoplasm and for the growth and survival of the tissues. Very few tissues can exist if their water content is as low as 10 per cent (Daubenmire, 1964). Water is essential for many physiological functions and it is most critical at certain periods in the life of a plant. For example, an adequate supply of water is essential for seed germination while too much water can hinder the ripening of a crop.

Plants have been classified into three main groups on the basis of their

SOIL-PLANT RELATIONS

tolerance to water. Hydrophytes are moisture-loving plants which grow normally with their roots in water or in damp soils such as swamps and bogs. Xerophytes, on the other hand, are plants which are able to withstand long periods of drought. The mechanisms for doing this include thin needle-like or waxy leaves to reduce the rate of transpiration and so the intake of water; and long tap roots which can draw up subterranean water from beyond the normal rooting zone of other plants. Other xerophytes escape adverse water conditions by growing annually and completing their life cycle within a very short period of time when soil moisture is adequate. Xerophytes are characteristic of arid and semi-arid areas.

The largest group of plants, however, are the mesophytes which show a wide range of adaptation to life in habitats which are neither too wet nor too dry. They are adapted to seasonal changes in soil and atmospheric moisture conditions. In the tropics where temperatures are constantly high and are more than adequate for plant growth, water is the master limiting physical factor in soil fertility. Most tropical plants and crops are mesophytic and tropical agriculture is adapted to the seasonal rhythm of soil moisture changes.

Soil water consists of three types in relation to its availability or suitability for rapid plant or crop growth. These are (1) superfluous, (2) desirably available, and (3) unavailable (see Chapter 2).

Superfluous water is the moisture in excess of that held at field capacity,

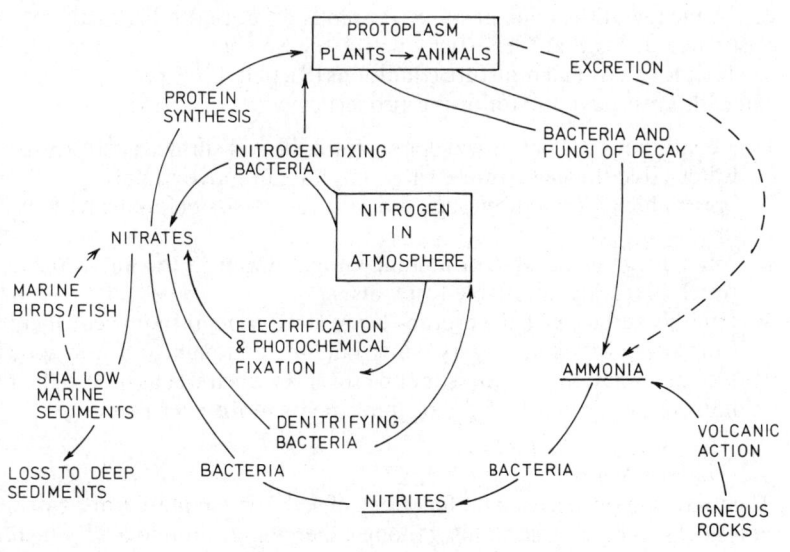

Fig. 9.1 *The nitrogen cycle*

usually the free-draining water in the soil. This kind of water leads to poor aeration and its attendant effects on chemical and biological elements and activities in the soil. Available water is that which is held in the soil between field capacity and the wilting coefficient. Vegetative growth is at its best when the soil moisture content is well above the wilting coefficient, because the rate of uptake of water by plants tends to decrease as the wilting coefficient is approached. Water that is held in the soil below the wilting coefficient or at the permanent wilting coefficient is referred to as unavailable water.

The control of soil moisture by man is very important. But unfortunately the soil properties, texture, and structure, which determine soil moisture relations as well as other physical factors of soil fertility, are very difficult to change by man. In controlling soil moisture, it is not only the physical properties of the soil that are important, but also the nature of the underlying parent material and local climatic conditions. Man has achieved a great measure of success in supplementing soil moisture content by irrigation. But he has been less successful in removing excess moisture from soils by drainage schemes (Cruickshank, 1972).

Soil Tilth
Tilth vaguely embraces the size-distribution, tenacity (or stability), and permeability (to air, water, and root penetration) of soil aggregates. Tilth is defined as coarse or fine depending on the size of the aggregates; and as mellow or raw depending on how easily the aggregates crumble when dry. A mellow tilth is one in which the clods do not crumble easily into dust when dry and are less sticky when wet. So soils with mellow tilth can be cultivated at all moisture conditions (Russell, 1966, p. 423). A good soil tilth must have the following properties:

1 a continuous system of wide pores from the soil surface down to the water table through which water and air can move rapidly;
2 pores which are stable enough to last several years before being filled up;
3 soil aggregates capable of holding enough water in the micro-pores which is readily accessible to plants;
4 crumbly surface soil, the crumbs being large enough to prevent them from being carried away by wind, but small enough to allow good seed germination; they must not be so sticky when wet as to lose their individuality when tractors or implements move over them.

Soil Organic Matter
The following properties and functions of soil organic matter are worth noting. As humus, it contains, among other things, humic acid which, in terms of quantity, is its most important constituent. These organic acids

are potent chelating agents forming complexes with metallic ions. The acids dissolve and mobilize organic compounds containing iron, magnesium, and aluminium complexes.

Humus contains organic polymers, i.e. fibrous or thread-like substances which help in binding soil particles together by means of strong hydrogen bonds. As such it is important in building a strong, stable soil structure which will facilitate the free movement of air and water. Furthermore, organic particles are hygroscopic, thereby augmenting the moisture-retentive capacity of soil. They also absorb heat and so influence soil temperature.

Humus is a store and supplier of plant nutrients, chiefly nitrogen, phosphorus, and sulphur, which are the growth-promoting elements in plants. The organic colloids have positive and negative exchange sites and so are able to hold and exchange anions and nutrient cations in the soil. In short, soil organic matter is of critical importance in maintaining soil fertility in soils which have a low clay content or in which the clay has low cation exchange capacity as in the highly weathered, kaolinitic soils

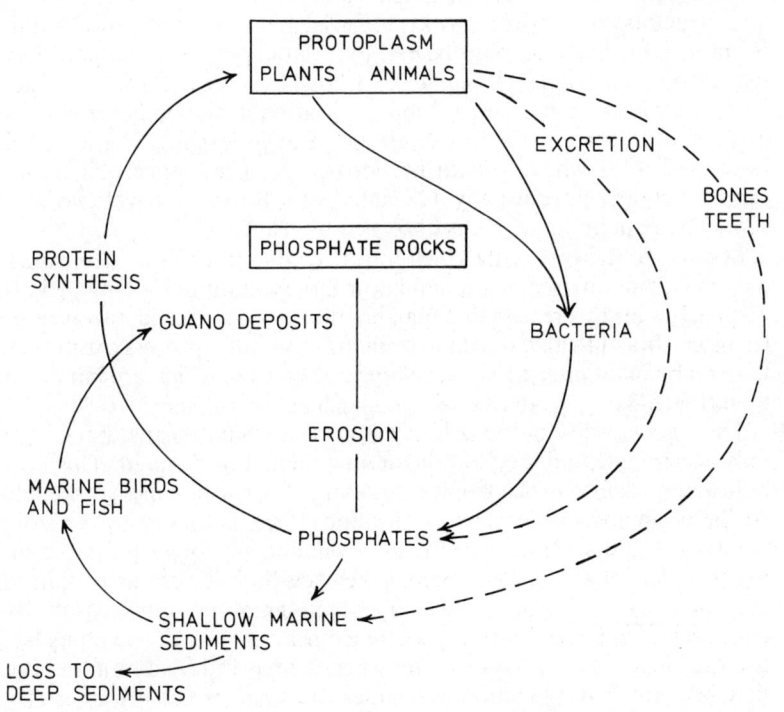

Fig. 9.2 The phosphorus cycle

of the tropics. Indeed, tropical agricultural systems depend very much on the break-down of litter and the mineralization of organic nutrients to restore and maintain the fertility of topsoils (see Nye and Greenland, 1960).

Nitrogen is clearly the most important organic nutrient in the soil. Plant need of nitrogen is by far greater than that of any other nutrient. Nitrogen cannot be used directly by plants but must first be converted by a complex process (referred to briefly in Chapter 4; see Fig. 9.1) into ammonia (NH_3^+) and nitrate (NO_3^-). Organic nitrogen will be oxidized by micro-organisms and mineralized into ammonia. This is augmented from the atmosphere by the direct fixation of nitrogen by free-living bacteria (e.g. Clostridium, Azotobacter), the symbiotic bacteria (e.g. Rhizobium found in nodules), and the blue-green algae. Some of the ammonia is converted to the ammonium ion (NH_4^+) which is held on the negative exchange sites of the soil clay-humus complex; part of it is taken up by micro-organisms and is locked up or immobilized in their bodies. But most of the ammonia is oxidized to nitrate in two stages in the process known as nitrification. The group of autotrophic bacteria known as Nitrosomonas first convert the ammonia to nitrite (NO_2^-) which is very toxic to plants. The nitrite is quickly oxidized to nitrate by a second group of autotrophic bacteria, Nitrobacter. Part of the nitrate produced is lost to the system through (1) leaching, (2) denitrification by denitrifying bacteria, the pseudomonas, and (3) immobilization in the bodies of micro-organisms. In general, the loss of nitrate through leaching is greatest in loose sandy soils, whereas denitrification occurs more in anaerobic conditions which are more usually associated with heavy, fine-textured clay soils. The remaining nitrate is absorbed by plants.

Because of the ease with which nitrate can be leached or denitrified, it is difficult for nitrogen to accumulate in large quantities for long periods of time. It is for this reason that man has developed and used nitrogenous fertilizers. Man has also sought to facilitate the natural process of nitrogen fixation by planting leguminous plants, the roots of which are inhabited by nitrogen-fixing symbiotic bacteria such as Rhizobium.

The susceptibility of the nitrate anion to leaching means that it will only accumulate under conditions of low rainfall or drought. This is of critical importance in the tropics, especially those with seasonal rainfall. At the beginning of the rains nitrification proceeds to add to the store of nitrate. But the nitrate accumulation will not stay long because leaching soon sets in as the rains become heavier. So it is very important in the tropics to plant crops at the very beginning of the rainy season. By contrast, in temperate latitudes, where temperature is the controlling factor, leaching is at its worst in winter when there is little evaporation. But then bacteria do not function well either and so not much nitrate is produced until fair weather arrives in the summer.

The nitrate anions absorbed by the plant go to form amino-acids from which proteins are made. Nitrogen is associated with the growth of tissue and the deep green colour of plants. In cultivated land, for example, there is a direct relationship between the amount of nitrogen in the soil and crop yield. If it is in short supply in the plant, the plant draws from the soil and its old parts to the growing parts, so that the old parts turn yellow and brown. Nitrogen serves as a regulator governing the utilization of potassium, phosphorus, and other nutrients by the plant. Plants not receiving sufficient amounts of nitrogen are stunted in growth and commonly have restricted root systems; while too much nitrogen leads to excessive vegetative growth which may delay maturation and expose the plant to attack by fungi because the cell walls are too thin.

Phosphorus is an important constituent of plant protoplasm. It enters the plant as the orthophosphate anion $(H_2PO_4^-)$ and its role in the plant is to store energy. It also controls the photosynthetic process in plants. Phosphorus exists in the organic form as chemical compounds and in the inorganic form as mineral salts.

The inorganic phosphorus pool is the bedrock. Weathering of the rocks leads to the formation of orthophosphate anions $(H_2PO_4^-)$ which are absorbed by plant roots (Fig. 9.2). Not all the inorganic phosphorus is immediately available to plants because some of it is taken up by micro-organisms and so becomes immobilized in their bodies. This competition from micro-organisms is often detrimental to plants. The phosphorus taken up by plants is returned through litter fall, and that passed on from plants to animals is returned in the excreta and the dead bodies of these animals. The litter is decomposed and the organic phosphorus is mineralized to the orthophosphate anion for recycling through the system.

The phosphorus reserve in the rocks is limited and is subject to great losses through weathering, leaching, and erosion without replenishment from the atmosphere. Consequently crops depend very much on organic phosphorus returned by plants and animals in the soil organic matter. Much of this is lost when the land is cleared for cultivation; phosphorus is thus about the most limiting nutrient in its effect on crop yield. Organic fertilizers which are rich in phosphorus, e.g. farm manure, guano, seaweeds, and bone meal, are commonly used to improve soil fertility. The phosphorus in these organic fertilizers is generally more soluble and available to plants than the inorganic phosphorus in mineral fertilizers.

Sulphur is only of secondary importance to plants compared with nitrogen, phosphorus, and some of the metallic cations. But it is nevertheless an essential growth-promoting element. It is used in the production of amino-acids which are the building 'bricks' of protein. It is absorbed by plants as the sulphate anion $(SO_4^=)$ in solution.

Like phosphorus, the major source of sulphur is the weathered bedrock where it exists in the form of inorganic compounds which are reduced

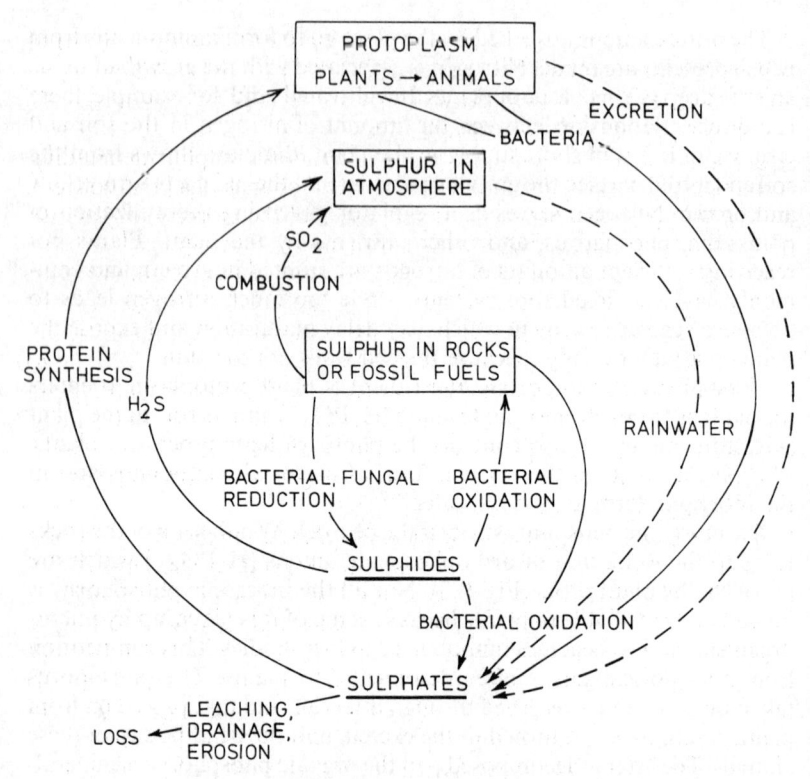

Fig. 9.3 The sulphur cycle

to sulphides under acid and anaerobic soil conditions. The sulphides are oxidized to sulphate notably by a group of autotrophic bacteria, *Thiobacillus thiooxidans*. The oxidation of the sulphides is never assured and the activities of certain bacteria even have the effect of reducing sulphates back to sulphides. Also, a considerable amount is lost to leaching, drainage water, runoff, and erosion. So the main sources of available sulphate in soils are (1) the soil organic matter, (2) the sulphur dioxide of the atmosphere, and (3) mineral (especially superphosphate) fertilizers. The sulphur in soil organic matter comes from plant residue and the excreta and dead bodies of animals (Fig. 9.3). Sulphur dioxide in the atmosphere is washed down by rainfall, this reserve being considerably augmented where there are industries burning coal, oil, or gas for fuel, or using sulphuric acid, or smelting sulphide ores. In coastal areas, the atmosphere receives a considerable amount of sulphur dioxide from the sea. Soils in coastal and industrial areas, therefore, tend to have an adequate supply of sulphur, while in non-industrial areas the amount in the soil decreases with distance from the sea. There is also a marked difference in the sulphur content of soils between arid and humid areas in non-industrial areas. Lack of rain in arid areas means that the atmospheric reserve is not even being washed down on to the soil.

Cruickshank (1972) has noted that the most sulphur-demanding crops—cotton, maize, groundnuts, rice, jute, bananas, tea, and coffee—are commonly grown in non-industrial areas of the tropics, so that sulphur deficiency in soils is becoming widespread. These areas can take a cue from the advanced agricultural economies of temperate latitudes by making use of mineral fertilizers. Among these are the superphosphates, ammonium phosphate, potash sulphate, and gypsum. The use of green and farm manures is also important. In these tropical areas large quantities of sulphur stored in the vegetation are volatilized during the annual bush fires. Parts of these are, however, washed back into the soil by rain.

The Exchangeable Bases (Metallic Cations)
These include calcium, potassium, magnesium, sodium, and ammonium, held on the negative exchange sites of the clay-humus complex in the soil. Of all the nutrient cations, calcium is usually the most abundant on the exchange sites. This is not surprising because calcite, the mineral element from which calcium is derived, occurs widely in very many rocks and especially in limestone. Calcium is, however, easily leached and replaced by H^+ ions on the exchange sites even in soils developed on limestone. Hence, as a result of its abundance in the soil and the ease with which it is leached, the calcium concentration is often used as a measure of the degree of leaching of the soil. Calcium is not stored in the soil as nitrogen, phosphorus, or sulphur in soil organic matter but it can be supplied by liming and the use of fertilizers such as superphosphates.

Calcium is needed in the soil to correct soil reaction (pH) and to promote plant growth. Plants which have an insufficient supply of calcium are stunted, this stunting being apparent in the leaves and roots. If the soil is inundated with calcium, however, the calcium will interfere with the supply of other nutrients to the plant. Calcium also helps in the bonding of soil particles to create a strong stable soil structure, a condition necessary for good soil drainage and aeration.

In terms of quantity, potassium is second only to nitrogen among the nutrients required by plants. It is important in photosynthesis, and plays a major part in the production of amino-acids and protein from ammonium and nitrate ions. If potassium is in short supply in the soil, plants are less resistant to stress and the leaves tend to wither sooner than would otherwise have been the case. Whether the amount of potassium in the soil will be adequate for plant needs or not, depends, more often than not, on the relative proportions of nitrogen and phosphorus in the soil. High concentrations of nitrogen and phosphorus may cause potassium deficiency but if the nitrogen and phosphorus levels are comparatively low, then the potassium in the soil may be sufficient for plant needs.

Potassium, like calcium, is derived mainly from mineral elements in rocks, being particularly common in feldspars, muscovite and biotite micas, and illite. It is more susceptible to leaching than phosphorus, and accumulates in the soil only in areas of low rainfall. Furthermore, like phosphorus, potassium can become fixed and unavailable to plants. However, it can be added to the soil through the application of mineral fertilizers, e.g. potassium chloride, sulphate, or carbonate. Crop and plant residues and manures add substantial quantities of potassium to the soil. In the tropics, when the forest is cleared and burnt for cultivation, potassium is the most abundant of the nutrients returned to the soil in the ash (Nye and Greenland, 1960).

Magnesium presents less problems than most other nutrients. It is not required in quite the same quantities as, say, calcium or potassium, and the concentration in the soil is less. But the reserve in rock minerals (micas, chlorite, etc.) is virtually enough for plant needs and more than balances losses by leaching, erosion, and crop removal. Nevertheless, in the highly weathered, leached, and kaolinitic tropical soils, magnesium deficiency is widespread. Potassium and ammonium ions usually replace magnesium on the negative exchange sites (see Chapter 3) and this ion exchange may cause an apparent magnesium deficiency in a soil. Magnesium is added to the soil in lime, mineral fertilizers, and farm manure.

Sodium is a trace element required in small amounts by certain plants. Sodium in large concentrations in the soil becomes toxic to plants. Large concentrations are found commonly in soils of arid environments (the solonetz) and of areas near the coast.

Problems of Soil Erosion and Pollution

Thus, soil fertility is determined by the interplay of soil physical, chemical, and biological properties. These must be considered in conjunction with the extraneous factors of climate, relief, slopes, rocks, and vegetation with which soil properties are closely linked. Soil is the major component of land on which man depends for growing his foodcrops. In many parts of the world, especially the less developed countries of Africa and Asia, there is great pressure on land due to increased population. The resources of the soil are being exploited without any meliorative feedbacks with the inevitable result that soil deterioration and erosion have set in. At the other extreme, in the advanced economies of Europe and America, men have sought to improve yields from the land through the extensive use of fertilizers and pesticides and insecticides. These have now led to soil pollution and general environmental deterioration in many areas. Clearly, man's activities in both the advanced and the less advanced economies pose serious problems for the maintenance of the balance of nature and the conservation of the environment. This matter is considered further in the final chapter of this book.

Chapter 10

SOIL EROSION AND CONSERVATION

Broadly defined, erosion is the wearing away or denudation of land masses. The term is derived from the Latin *erodere* which means to gnaw away. Although when applied strictly it refers only to the acquisition or detachment of material by erosive agents, geologists in particular and earth scientists in general apply it more broadly to include both the process of material acquisition or detachment and that of transportation of such materials by the geologic agents of water, wind, and ice. Accordingly, soil erosion can be defined as the processes of detachment and transportation of soil materials by the erosive agents of water, wind, and ice.

Soil erosion is commonly classified as one of two types. First, there is the normal geological erosion, which operates on the earth's surface wherever energy flows, whether as water, wind, or ice. The second type of soil erosion is accelerated soil erosion, caused mainly by the activities of man which upset the balance between soil, vegetation cover, and the erosive power of the various geological agents. This is the type which, because of its destructive nature, has been attracting man's attention. It is also the type which is usually seen as soil erosion proper. It started much later than normal geological erosion, dating back only to the time of man's settled existence, specifically to the beginning of organized agriculture. It is possible to imagine that it started when the early nomadic pastoralists first upset the delicate balance between plant cover and the erosive force. It has since been intensified and worsened by crop cultivation and animal grazing, among other causes.

The seriousness of accelerated soil erosion was felt as soon as it started, so that records of man's efforts to find a lasting solution to the problem actually go back thousands of years. This is understandable at least from the known and conspicuous consequences. These include the physical loss of soil constituents leading to severe economic loss arising from reduced yield, total crop failure, and/or wasted efforts and money on unsuccessful soil conservation projects; total loss of vast areas to gully and other badland phenomena; accelerated sediment removal leading to increased pollution of stream, sea water, and the atmosphere; silting of lakes both natural and man-made; damage to irrigation canals; etc. This chapter summarizes the present state of our knowledge about both soil erosion and soil conservation.

The Nature of Soil Erosion

Soil erosion involves two separate but sequential events. First, soil particles are detached (or torn loose) from the soil mass; this is then followed by the transportation of the detached materials from their original locations. The deposition of such materials, although of interest to soil science generally, is usually not considered as part of soil erosion. In effect, therefore, the erosive capacity of any erosional agent will be a function of the two independent variables, expressed as (a) the detaching (or dislodgement) capacity and (b) the transporting capacity of the agent.

The erosive capacity of an agent is itself controlled by (i) the amount of abrasive fraction contained in the flow and (ii) the nature of the soil being eroded. An experiment performed by Ellison (1947) has shown that the detaching capacity of surface water increases, while the transporting capacity decreases, as the abrasive-fraction content increases. Maximum erosion will therefore be expected to occur when the detaching and the transporting capacities of the agent are well balanced; i.e. when the flow contains just enough abrasive materials to detach as much as the flow can carry. In relation to the nature of the soil itself, the aspects that are important are its texture and structure. While detachability increases with increasing soil particle-size, transportability does the opposite. Thus, while it is more difficult for, say, water to detach clay than to detach sand, it is a lot easier to transport clay than sand. By implication, erosion will be more on medium-sized soils (silt) than on either coarse-grained or fine-grained soils; while, depending upon which of the indices of erodibility we are considering, clay or sand may be of low or high erodibility. We shall have more to say on this topic later in this chapter.

Soil erosion is not necessarily a bad thing. Apart from being a natural process which, as part of the hydrological-cum-rock cycle, represents a natural phenomenon, it also provides material for further soil formation. Indeed, alluvial soils, loess soils, and glacial tills are among the most productive soils in the world. There is a need for certain elements to cycle in order to maintain their existence. This is precisely what the processes of weathering, erosion, deposition or sedimentation, cementation, metamorphism, and vulcanism do. The process of the rock cycle will stop or will take a different course, if soil erosion does stop.

Finally, soil formation is a continuous process. The removal of surface soils facilitates the formation of a new one to replace the lost one in a dynamic process. Indeed soil formation and soil erosion operate *pari passu*, and as long as the rate of soil erosion is not faster than that of soil formation, there is no trouble whatsoever. This is what obtains in many terrestrial (natural or virgin) ecosystems, where the vegetation cover maintains a sort of balance between soil formation and soil destruction. The problem starts only when this balance is upset. This introduces a third factor to the two considered earlier. As man uses the soil, the severity of the resulting erosion will depend on (i) the nature of the soil; (ii) the nature of the removal agents; and (iii) the extent of interference with both the vegetal cover and the soil surface. These are the factors of soil erosion which will be considered next.

The Factors of Soil Erosion

The factors which are commonly considered in connection with soil erosion are the nature of the soil and soil surface, whether compacted or loose; the nature of the land surface features or physiography; the climate; vegetation cover; and the extent of human interference. Nature maintains a system or systems of natural checks and counter-checks so that, as noted above, erosion is minimized, the control key being vegetation cover. But although the result of human interference can be decisive in some cases, the other factors also exercise strong controls. Indeed, it is difficult to separate the effects of one set of factors from those of the others since they are all intimately linked. But for the sake of simplicity, we shall treat each factor separately, grouping them into physical and human factors.

Physical Factors
The susceptibility of a soil to erosion, particularly geologic erosion, is essentially a function of the physical factors of the area. These include (1) the lithology, structural stability, infiltration and permeability rates, porosity, or water-holding capacity of the soil; (2) the prevailing climate which among other things controls both the type and the intensity (capacity) of the affective erosion agent or agents, whether water, wind, and/or ice; (3) the physiographic factor or surface configuration, particularly slope, relief, and drainage characteristics; and (4) the amount and type of vegetation cover. Each of these factors is a complex in itself, to the extent that each is controlled by, and is made up of, a number of variables or factors, while they together exhibit a system of links and interlinks.

THE NATURE OF THE SOIL
The attributes of the soil which influence its erosive capacity are its degree of detachability and transportability; its infiltration and permeability rates; the extent of its aggregation and surface sealing; its total depth or, more important, that of the solum; and its water-holding capacity; to list only a few important ones. These factors are sometimes summed up in the expression 'structural stability', which describes the resistance or otherwise of a soil to the various destructive forces. Soils which break down and get washed away readily are said to be of low resistance or

to be structurally unstable, while those which do not break down readily and are difficult to remove are structurally stable. The concept of structural stability, useful as it is for a general discussion, is rather complex. Because it involves all the physical properties as well as the composition of the soil, its measurement is not simple. It also has a dynamic attribute to it. The method of describing the various aspects separately is therefore adopted here.

By far the most important agent of soil erosion is water. Soil erosion by water, however, does not occur without runoff, which is intimately related to the rainfall intensity, the infiltration capacity and permeability characteristics of the soil, the nature of aggregation and surface sealing, and similar factors.

The rate of infiltration of dry soils is very rapid for a short time. As the soil becomes wet, the infiltration rate decreases rapidly until finally it reaches an equilibrium rate. The relationship is often expressed empirically by the equation

$$Y = At^B \qquad \qquad (10.1)$$

where Y is the cumulative infiltration in inches or centimetres, t is time in minutes, and A and B are constants which depend on soil characteristics. The exponent B varies between 0 and 1·0. The equilibrium rate is principally a function of soil texture and structure as well as the wetness of the soil prior to a rainfall. Other factors are surface sealing by rainfall impact and vegetation cover. The nature of the surface, whether dug up or not, is also important; so also is the porosity or water-holding capacity and the depth and nature of the soil's underlying horizons.

Soil structure or aggregation is mainly dependent on organic matter content and the type of bases in the soil. Soils devoid of organic matter have very unstable aggregates. By contrast, a soil high in organic matter is more stable. Similarly, divalent cations (e.g. calcium) increase aggregation while monovalent cations (sodium) decrease aggregation and disperse the soil. Also the rate of surface sealing during rainfall is increased by organic matter content, resulting in reduced infiltration, increased runoff, and increased erodibility. Another factor is the depth of the topsoil; deep topsoil allows water infiltration to proceed unrestricted for a time until layers of varying permeability and porosity are reached.

CLIMATE

There are two main ways in which climate influences soil erosion. First, it dictates whether the erosive agent will be water, wind, or ice. In warm and wet areas, water is the dominant erosive agent. Wind is powerful in desert, especially warm desert, regions, while ice is dominant in cold regions. Secondly, climate regulates the intensity of soil erosion. This is

```
                    ┌──────────────┐
                    │ WATER EROSION│
                    └──────┬───────┘
                       Affected by
           ┌───────────────┴───────────────┐
┌──────────────────────┐         ┌──────────────────────┐
│ Dispersive effects   │         │ Susceptibility of    │
│ of raindrops;        │         │ soil to dispersion   │
│ amount and           │         │ and movement         │
│ velocity of runoff   │         │                      │
└──────────┬───────────┘         └──────────┬───────────┘
      Determined by                    Determined by
```

Fig. 10.1 *Factors of soil erosion by water*

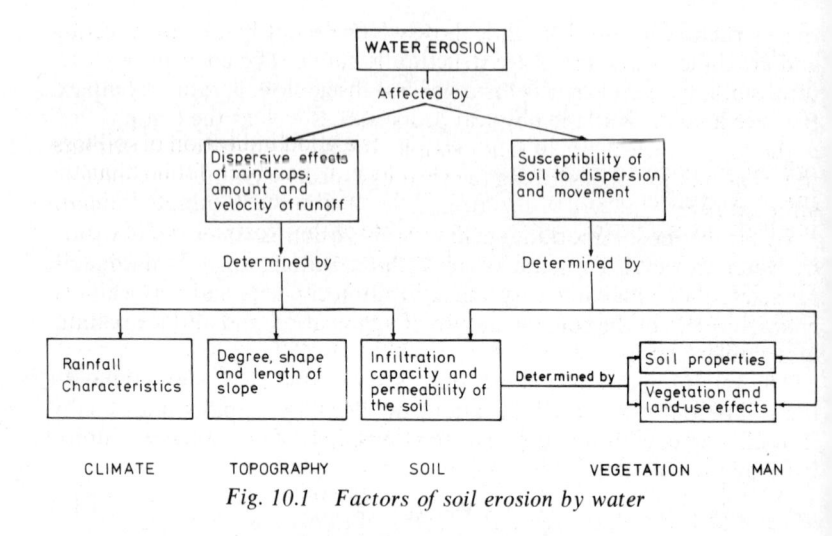

particularly true of water erosion, where the quantity, intensity, energy, and distribution of rainfall are important.

Up to a certain point, erosive capacity increases with increasing rainfall. The trend changes with the onset of vegetation, which counters the effects of increasing rainfall and produces a marked decrease in the net erosion rate. In effect water erosion is theoretically most effective in the semi-arid areas (Fig. 10.1).

Perhaps more important than mean annual rainfall is the rainfall intensity. Soil erosion requires energy, which is provided by rainfall. The equation which relates rainfall intensity to kinetic energy is

$$Y = 916 + 331 \log_{10} X \quad \ldots \ldots \quad (10.2)$$

where Y = kinetic energy in feet-tons per acre-inch and X = rainfall intensity in inches per hour. Another form of the same equation is

$$Y = 206 + 87 \log_{10} X \quad \ldots \ldots \quad (10.3)$$

where Y = kinetic energy in joules per centimetre per square metre and X = rainfall intensity in centimetres per hour. The kinetic energy of individual rainstorms can thus be computed if rain-gauge records are available.

The energy for water erosion is provided principally by the potential energy of raindrops. In other words, the mean raindrop size determines the amount of energy an individual rainstorm will impart. Increase in the size and number of individual raindrops means increase in rainfall intensity, kinetic energy, and consequently erosive capacity. The relationship between drop-size and intensity is expressed by the equation

$$D_{50} = 2\cdot23 I^{0\cdot182} \quad \ldots \quad (10.4)$$

where D_{50} is the median drop diameter in millimetres and I is the rainfall intensity in inches per hour.

But while the energy of a rainstorm gives a good indication of soil loss, other factors come into play. For instance, it has been found that the interaction effect of kinetic energy and the maximum 30-minute (rainfall) intensity gives a better (very good) measure of the erosive force of a rainstorm. The measure is called the energy-intensity product or 'EI' which is an interaction value, or the product of the kinetic energy of a rainstorm expressed in hundreds of foot-tons per acre (or joules per centimetre per square metre) and its maximum 30-minute intensity in inches per hour. These values can also be computed from reliable rainfall records. The sum of individual storm (EI) values for a given time provides a numerical evaluation of the erosivity of the rainstorm within the period. Annual figures of (EI) are commonly referred to as rainfall erosion indices.

Apart from vegetation and the nature of the soil, temperature is another factor which influences the effectiveness of rainfall as an erosion agent. Temperature affects the type and amount of vegetation cover; the rate of decomposition both of rocks and of organic matter; and the rate of water loss through evaporation and evapotranspiration.

PHYSIOGRAPHY

The attributes of physiography which affect soil erosion include slope steepness, length and curvature; topographic or surface roughness; and drainage network, to list only a few important ones. Among the first comprehensive studies of the effect of slope on soil loss was Zingg's (1940) work, which among other things found that soil loss varies as the 1·4 power of the per cent slope. This means that an exponential equation expresses the relationship between soil loss and slope steepness. Zingg's work was however in error as it applies to flat surfaces, where zero soil loss is suggested. Later workers have therefore proposed alternative equations. One such equation is

$$R = 0\cdot10 + 0\cdot21 S^{4/3} \quad \ldots \quad (10.5)$$

where R is relative soil loss in relation to unity loss for a slope of 3 per cent and S is per cent slope (Smith and Whitt, 1947). The problem of soil erosion (by water) is generally not serious on gently sloping to flat terrains, where sedimentation rather than erosion may be the problem. However, as slope angle increases, erosion increases, particularly in the case of sandy soils. Here, slope steepness increases the gravitational pull of loose soil downslope. Slope steepness also increases the velocity of runoff, which can then pick up more soil for transportation downslope. Slope steepness is of little consequence for wind erosion, which is less dependent on gravity.

The relationship between soil loss and slope length has also been investigated. Zingg (1940) observed that total soil loss varied as the 1·6 power of slope length, while soil loss per unit area varied as the 0·6 power of the slope length. The same relationship has been expressed as approximately equal to the square root of the slope length ($L^{0.5}$) for soils on which runoff rate is not affected by slope length, and more than 0·5 on soils where runoff increases with slope length. The complex influence of vegetation cover, rainfall intensity, slope steepness, surface roughness and infiltration makes the effect of slope length difficult to isolate. Slope length is commonly measured from the point where overland flow starts to the point where deposition begins or where runoff enters a well-defined channel.

The form or shape of a slope also affects soil erosion. Three shapes—convex, straight or rectilinear, and concave—are known and each influences soil erosion differently. Convex slopes increase in steepness towards the foothill or footslope. This means that runoff velocity also increases downslope and with it soil erosion. By contrast, concave slopes flatten out downslope, leading to decreased velocity and deposition of sediment. With high-intensity storms, however, water tends to concentrate on concave slopes and gulling may be initiated. The effect of slope is constant on straight or rectilinear slopes.

Land surface roughness generally slows down the speed of flow of energy and so reduces erosion. It may in fact encourage deposition. Roughness of surface is particularly effective in the case of wind erosion, where alternate ridges and depressions trap saltating particles and thus stop the normal build-up of eroding material downwind. Also in the case of water erosion the roughness factor can be very important, since it exerts a very strong influence on velocity.

The surface drainage characteristics of an area also affect the rate of soil erosion. Expressed either as drainage density (Dd), as stream frequency (Fs), or as both, i.e. as drainage intensity (Id), there is some link between the drainage texture and removal of surface material. A study conducted in a part of Australia shows, for instance, that the relationship between drainage intensity and volume of removal may be expressed by the equation

$$Y = 1.05X + 446 \quad \ldots \ldots \quad (10.6)$$

where Y = volume of removal in cubic kilometres and X is drainage intensity.

VEGETATION
Finally, there is the factor of vegetation cover and land use. This is perhaps one of the best-studied factors of soil erosion. It is still being studied in many parts of the world in what are called runoff plots, undertaken mainly for agricultural experiments. The runoff plots at the International Institute of Tropical Agriculture (IITA) in Ibadan and at the Institute of Agricultural Research, Samaru, Zaria, are good examples (Table 10.1). Forests, especially the natural ones, are very effective in controlling erosion. The tree canopy intercepts rainfall and reduces its energy. Drops that reach the ground surface are quickly taken up in the leaf litter and from there into the generally highly porous forest soils. Thus a number of observers have reported no runoff from plots made in natural forest. Therefore, with the considerably reduced rainfall impact and little or no runoff, soil erosion is at the minimum in forest-covered areas. Similar effects result in the case of wind erosion. However, if this forest is disturbed in any way, e.g. by logging, bushfire, etc., the natural protection of the soil may be destroyed. Finally, forest plantations are like row crop and so do not offer the same kind of protection for the soil as natural forests. This has been confirmed by gullying action initiated in many forest plantations, one example of which has been recently described (Faniran and Areola, 1974). In short, the best protection against soil erosion is natural forest and the clearing of forests or their disturbance in any form has always meant accelerated soil erosion. This brings us to the role of man in causing or accelerating soil erosion.

The Human Factor
The human factor in soil erosion is made manifest through man's activities, including farming, grazing, mining, and civil engineering constructions. Each of these activities disturbs the delicate balance of terrestrial ecosystems and may induce accelerated erosion. The effect is most quickly felt and is also most serious in marginal areas, whether in terms of vegetation cover or population density. Thus in semi-arid regions, the scanty vegetation cover is readily destroyed by overgrazing and primitive agriculture, leading to the exposure of soils to the erosive forces, both of wind and of water. Similarly, where population exceeds the carrying capacity of the land, land deterioration results. Let us see what some of these activities of man are and how they induce, encourage, or accelerate soil erosion.

AGRICULTURE AND SOIL EROSION
This is perhaps the most important cause of soil erosion, principally because of its widespread occurrence. It may take the form of crop cultivation or grazing. In the case of crop cultivation, the extent of soil erosion

TABLE 10.1
Effect of Slope and Cultural Practices on Soil and Water Loss (Ibadan IITA, 1972)

Total rainfall = 104·02 cm

Rotation	10% Slope		5% Slope		10% Slope		15% Slope	
	Soil loss (tons/ha)	Runoff (cm)	Soil loss (tons/ha)	Runoff (cm)	Soil loss (tons/ha)	Runoff (cm)	Soil loss (tons/ha)	Runoff (cm)
Bare fallow	3·2734	21·0128	37·6269	22·2330	50·3317	22·7822	115·1102	17·6752
Continuous maize (conventional tillage)	1·068	7·4728	2·1845	14·4857	5·7672	8·6172	13·4638	7·3206
Continuous maize (conventional tillage + mulch)	0·000	0·052	0·1086	2·7732	0·1402	1·8825	0·1822	4·5216
Maize–cowpeas (conventional tillage)	0·3503	1·6767	1·9454	6·0309	2·2070	3·4483	5·8074	4·013
Maize–cowpeas (zero tillage)	0·3050	2·4201	1·5315	10·2471	4·8528	10·6463	8·2758	9·3704

From IITA. Ibadan, *Annual Report*, 1972.

will depend upon the system in vogue, the intensity of cultivation, and the type of crop grown, among other things. The highest possible amount of erosion occurs on bare or clean-tilled fallow land, especially if ploughed at the beginning of the fallow period. The amount of erosion on such a piece of land depends on previous cropping, whether or not residues are retained or removed, and on the amount of surface detention left by primary and secondary tillage. Soil erosion is generally less serious on soils fallowed after several years of perennial grass cover than on soils in continuous maize or cotton cover with all residues removed. Similarly, a field ploughed, disked, and harrowed is more susceptible to erosion by the first few rains of high intensity than a rough-ploughed field.

Opinion is somewhat divided on the real implications of the so-called primitive agriculture—e.g. shifting cultivation and/or bush fallowing—for soil erosion. While some people think it is destructive of the soil, others suggest that it is well adapted to the environment. The critical issues here seem to be the length of the fallow period and the cropping system. Where mixed cropping is prevalent, as it is in many developing countries where between two and five crops including cover crops are grown together on a farm plot, soil exposure is minimized and with it the degree of soil erosion. Similarly where the fallow period is long enough for complete soil recuperation, the cumulative effect of soil erosion is considerably reduced. However, where farm plots are clean-tilled and/or where the fallow period is short and the cultivation period long, soil deterioration is rapid, leading to depleted vegetation cover; consequently accelerated erosion ensues.

This is the situation in many parts of the developing world today where population has increased so rapidly as to cause serious land shortage and pressure. Some examples from Nigeria are described below. In the meantime, however, it may be observed that, given the type of farming instruments in use, e.g. hoes and cutlasses, the widespread practice of mixed farming and the generally short cultivation and long fallow periods in some countries, the bush fallow system seems to minimize rather than accelerate soil erosion. It is only when these practices are changed, e.g. by the use of heavy machinery, growing of single crops, prolonged cultivation period, and/or shortened fallow periods, etc., that trouble starts.

This brings us to the issue of the proper method of land management. The management factor in soil erosion measures the deviation in the field application of soil, crop, and conservation practices from the standards which exist in experimental plots. This includes the timing of operations, the handling of crop residues, the application of fertilizers, precision in the application of conservation practices, and the use of optimum seedbed (crop density) operations. If conservation practices are strictly adhered to, soil erosion will be negligible. The bane of some forms of 'primitive' agriculture is that very few standard scientific conservation practices exist. There is no way of knowing how much soil is actually being lost per unit time, let alone checking such loss. A high premium is placed on nature and natural regeneration, except in a few places where manuring is practised.

MINING

Mining is also very destructive of land, especially if it is open cast mining. Destruction to soil by mining occurs in two main ways. First, it renders vast areas valueless for other land uses, except after reclamation. Secondly, and following directly from the first point, the resulting land shortage leads to pressure on the land, land exhaustion, and excessive soil erosion.

The Jos Plateau area of Nigeria provides a good example. This is one of the most intensively mined parts of the country, being the main source of the country's tin and associated mineral deposits. Mining started about 1904 and by now vast areas have been dug up. The situation is made worse by the method which requires the removal of large volumes of overburden. Consequently, approximately half of the Jos Plateau minefields were rendered useless for farming by 1950. This has meant great pressure on the remaining farmlands, leading to land exhaustion. The Jos Plateau is today one of the worst affected areas in Nigeria, as far as erosion is concerned (Grove, 1952).

ROADS, FOOTPATHS, SETTLEMENTS, ETC.

Footpaths, roads, and other excavations by man, apart from stripping the soil of its vegetation cover, also concentrate runoff and so initiate concentrated soil erosion, including, in suitable locations, the development of wide gullies. Similarly settlements, market places, and so on accelerate soil erosion in a way similar to roads and excavations. The role of these phenomena in gully development in parts of Nigeria's Anambra State has been described by many workers including Grove (1951) and Ofomata (1965).

Finally, let us consider very briefly the attempts being made to integrate all these factors to produce an index of soil loss. All the work done so far uses empirical equations of the kind

$$A = C \times S \times L \times K \times M \times P . \quad . \quad . \quad . \quad . \quad (10.7a)$$

or simply

$$A = CSLKMP . \quad . \quad . \quad . \quad . \quad . \quad . \quad . \quad . \quad (10.7b)$$

(Smith and Wischmeier, 1957, p. 895), where A is average annual field soil loss in tons per acre, C is average annual plot soil loss in tons per acre for a selected rotation with farming up- and downslope, S is the per cent slope and L slope length, both adjusted to give unity loss on a 3 per cent slope 90 feet long, K is a soil factor, P is the factor for conservation practices in relation to a unity value for up- and downhill farming, and M is the management factor, with management of the crop rotation plots taken as unity. A similar equation which is preferred by the US Department of Agriculture for estimating erosion by water or rainfall is

$$A = RKLSCP \qquad (10.8)$$

where A is average annual soil loss in tons per acre, predicted by the equation, R is the rainfall factor, K the soil erodability factor, L the length of slope factor, S the steepness of slope factor, C the cropping and management factor, and P the supporting conservation practice factor (terracing, trip-cropping, contouring, etc.).

In equation 10.8, the various factors are defined as follows. The rainfall factor (R) is a numerical value which expresses the capacity of the locally expected rainfall to erode soil from an unprotected (fallow) field. The value that is most commonly used here is the EI value already described above. This value has been shown to explain from 72 to 97 per cent of the variations in individual storm losses from tilled and continuous fallow soils in the American states of Missourri, Iowa, Wisconsin, Ohio, New York, and South Carolina.

The soil erodibility factor (K) reflects the erodibility of soils with varying physical characteristics. The factor represents or gives tons of soil loss per acre per unit of rainfall erosion for a slope of 9 per cent and length of 22·1 m (72·6 feet). The arbitrary choice of 9 per cent and 22·1 m (72·6 feet) for slope steepness and length respectively is not very satisfactory, but reflects the extent of the problem involved in the calculation. The procedure followed in the US is to rate some of the major soils, beginning with the least erosive, e.g. the deep, permeable, coarse soils, and proceeding to the most erosive, e.g. shallow sandy soils over impermeable material. About five to ten factors are found to be adequate to cover all the soils in a given state or country.

The slope length (L) and steepness (S) factors are usually combined. The equation used is

$$SL = \frac{L}{100}(0.76 + 0.53S + 0.76S^2) \qquad (10.9)$$

or

$$SL = \frac{L}{100}(1.36 + 0.97S + 0.138S^2) \qquad (10.10)$$

where S is percentage slope and L the slope length in feet. SL is a ratio of soil loss from a given percentage and length of slope to the soil loss from a 9 per cent slope 22·1 m (72·6 feet) long. Like the soil erodibility

(K) factor, the slope factor (SL) is not yet considered entirely satisfactory, and research is still continuing in both these directions.

The cropping-management factor (C) is the expected ratio of soil loss from land cropped under specified conditions to corresponding soil loss from continuous fallow. The influence of crops and cropping practice on erosion is affected by the kind of crop, the quality of the cover and root growth, water use by growing plants, quantity of prior-crop residues, etc. It is also possible to divide crop growth into stages or periods, based on relative uniformity of cover and residue, e.g. the rough fallow, seedbed, establishment, growing-crop, and residue or stubble stages. All these are considered in obtaining the C factor. The last factor, the conservation practices factor (P), takes into account the effect of contouring, terracing, and strip cropping.

Agents of Soil Erosion

Soil erosion is effected by two main agents: water and wind. In addition gravity or mass-wasting and ice destroy soils, but the areas where these operate are rarely agricultural lands. They are therefore not usually considered in any detail. For instance, mass-wasting, especially landslides and landslips, is common in badly gullied areas or otherwise unstable soil regions, which are rarely cultivated. Similarly, ice erosion occurs in cold regions, where the soil is not only frozen but also ice-covered to preclude cultivation. We shall therefore limit our discussion to water and wind.

Soil Erosion by Water

This is by far the most common type of soil erosion. Throughout the world, except in the true deserts and the ice-capped polar regions, the land surface is subject to degradation by water, especially if the surface is exposed to some extent.

Soil erosion is all the more serious because it operates in those areas where agriculture—whether arable or pastoral—is possible; that is, where it rains. Moreover, soil erosion by water affects the topsoils, which contain the bulk of the plant food. Thus, unless we understand the process of soil erosion by water, so that we can check it effectively or minimize its impact considerably, agriculture will be in great jeopardy from falling raindrops and flowing water.

The energy of falling raindrops is applied from above, and their main function in soil erosion is to detach soil particles from the soil mass. By contrast, the energy of flowing water is usually applied parallel with the surface; its main function in soil erosion is to transport (remove) soil materials to new environments. These two agents produce widely different effects on the soil, leading to a further subdivision of the erosion process into the raindrop type and the running water or flowing water type.

RAINDROPS AND SOIL EROSION

Soil erosion by raindrops is usually described as splash erosion. This is because when falling raindrops strike the ground surface or the thin films of water covering it, the soil particles fly in all directions. These are the particles which are detached from the soil mass on impact. The quantity of the detached soil is proportional to the detaching capacity of the drops and the detachability of the soil. This can be expressed empirically as

$$D_t = D_r \times D_s \qquad (10.11)$$

where D_t is total soil detachment hazard or total quantity of detached soil, D_r is the detaching capacity of raindrops, and D_s is the detachability of the soil.

The kinetic energy of the falling drop determines the force of the blow that must be absorbed at each point, while the horizontal area (size) of the drop determines the amount of soil particles that must sustain that blow. The kinetic energy of falling raindrops amounts to 10^4 ergs for drops of 2 mm radius, while a raindrop of 2·5 mm radius is sufficient to raise by 1 cm a body weighing 46 g. This energy increases as the 1·2 power of rainfall intensity. The splash erosion process is thus affected among other things by variations in the size of the raindrops, the drop velocity, and the rainfall intensity. Ellison (1947), who investigated these factors on silt loam, found that soil splash varied as the 4·33 power of the drop velocity, 1·07 power of the drop diameter, and 0·65 power of the rainfall intensity. Another study using rain applicators showed that as the drop size increased from 1 to 5 mm in diameter, erosion losses increased by up to 1,200 per cent.

On level surfaces, the splashed material tends to scatter uniformly over the surface in all directions, particularly when the raindrops have vertical drop. By contrast, when raindrops strike a sloping land surface, the major portion of the splash moves downhill. The result is more soil erosion on steep than on gentle slopes.

The factor of soil detachability (D_s) involves two principal groups of factors. The first group comprises those factors which affect the soil consolidation under impact of raindrops. If the soil consolidates through compaction and cementation under raindrop impacts the rate of soil detachment decreases. The second group of factors includes the stone fractions, firm clods, and other materials which resist breakage. The more of these, the less the rate of soil detachability. This factor has already been discussed. So also has the factor of vegetation or crop cover. This last factor is important in that it influences equation 10.11 which applies only to bare surfaces. The factor R indicates the effectiveness of vegetation or other covers in checking falling raindrops, or the impact conditional factor. The equation which includes this factor is

$$D_t = \frac{D_r \times D_s}{R} \qquad \qquad (10.12)$$

where R is the impact-conditioning factor and the others are as in equation 10.11.

Splash erosion involves a number of hazards. First, detachment means ready entrainment; that is, when a soil is detached, it is readily available for transport. Secondly, splash erosion involves the process of elutriation, or the washing out of the more valuable parts of the soil, leaving the stone and sand fractions in the field. This can be particularly serious in areas with some slope. Finally, splash erosion breaks down soil clods and crumbs and causes many of the aggregates to release their humus and other light materials, invariably for removal as suspended load.

SOIL EROSION BY SURFACE FLOW

Surface or overland flow occurs when precipitation exceeds infiltration and throughflow, that is, when the rate at which water reaches the ground surface is more than that at which the water is lost through soaking into the soil. This happens (a) at the base of slope adjacent to water channels; (b) in hollows and depressions; (c) at slope profile concavities; and (d) in areas of thin or low soil permeability.

There are two main types of surface flow—laminar and turbulent flow. Either may be as sheet flow or channel flow. In laminar flow: (a) the flow velocity is very slow, of the order of 1 mm/sec.; (b) parallel layers of water shear one over the other; (c) the shearing stress (Y) is proportional to fluid resistance to movement, i.e.

$$Y = U \frac{dv}{dy} \qquad \qquad (10.13)$$

where dv/dy is change in velocity from one layer to the next, Y is shearing stress, and U is a constant; (d) the layer of maximum velocity is below the water surface while velocity zero is at the boundary with bed or channel walls (if channellized flow); and (e) the paths of fluid movements are theoretically parallel and unmixed. Consequently, whether as sheet flow or as channel flow, laminar flow does not cause soil erosion, since it cannot support solid particles in suspension. It contains, however, soluble substances. This type of flow is not common on the earth's surface.

By contrast, in turbulent flow, velocity exceeds a critical value, so that the flow is characterized by a variety of chaotic movements, with secondary heterogeneous eddies superimposed at the main forward flow.

There is sufficient energy in this type of flow for transportation of solid particles whether as tractive load or perhaps more importantly as suspension load. This is the main cause of soil erosion, two types of which are known, namely, sheet erosion and channel erosion.

SHEET AND MICRO-CHANNEL EROSION

Sheet erosion is the more or less even removal of thin layers, or 'sheets' of soil from a sloping land. It is a rather inconspicuous type of erosion because the total amount removed in any storm is usually small; however, the cumulative effect can be very disastrous, if not checked. Sheet flow is itself rather ineffective as an erosive force, but when combined with raindrop effect, soil loss or erosion occurs. Indeed, practically all the so-called sheet erosion is actually splash erosion. Sheet erosion therefore occupies an intermediate position between raindrop and flow forces and it can be treated as part of the one or the other.

The first sign of sheet erosion is the exposure of light-coloured soil particles, following the removal of the organic-matter-rich surface layers. The two basic processes of soil particle detachment by raindrops and transportation by flowing water operate. Ordinarily, the sheet flow is incapable of doing these two jobs, but raindrops increase turbulence which then provides the required energy. But the detaching power and transporting power of sheet flow are functions of the depth and velocity of runoff for a given size, shape, and density of soil particles or aggregates. Studies made so far indicate that maximum soil loss occurs when the depth of flow is about equal to the mean diameter of the soil particles.

Sheet flow occurs mainly on smooth, uniform slopes which rarely exist, especially on cultivated fields, the surface of which are often irregular. On such irregular surfaces, rainwater accumulates in depressions, causing it to move into micro-channels or rills.

The boundary between sheet and rill flow is difficult to draw, but rills form as soon as surface flow begins. They vary in size from minute channels to a size that may just be easily observed; they may also join together to look like sheet flow. In such a flow, rill erosion is concentrated in the micro-channels while sheet erosion occurs in between them. Detachment and transportation of soil particles are both greater in rill than in sheet erosion, due to the greater velocity of the channellized flow. We may in this connection recall the following findings, that: (a) the amount of soil particle detachment by moving water is proportional to the square of its velocity; and (b) if the discharge of a flow is doubled, the erosive capacity increases five times while the competence, or the calibre of the particles it can transport, increases six times; (c) finally, rill erosion cuts deeper than sheet erosion, at times cutting into the subsoil.

CHANNEL FLOW OR GULLY EROSION

Farther downslope at favourable locations, rills or micro-channels join to make larger channels, with greater discharge and greater erosive powers. At this stage rill erosion changes into gully erosion, which is channel erosion that cuts so deeply into the soil that the surface cannot be smoothed out by ordinary tillage tools.

The rate and extent of gully development are closely related to the velocity of the runoff, which itself is closely related to the size of the drainage basin and its hydrological characteristics. Rainfall and soil characteristics, land slope, and vegetation are other important factors, all of which have been mentioned briefly in the general discussion at the beginning of this chapter. Of these factors, the rainfall characteristics, the nature of the soil, the land slope, and the extent or degree of human interference are decisive. Among other things, (i) the rainfall needs to be high enough to supply the large quantities of water needed to provide the energy for both detachment and transportation of the soil material; (ii) the soil needs to be deep, loose and light—e.g. the unconsolidated sandstones of the Enugu area or the thick alluvial (and colluvial) deposits of the Zaria area, both in Nigeria; (iii) the slope needs to be steep enough to encourage fast-moving runoff; and (iv) the vegetation cover needs to be cleared or otherwise depleted to expose the soil to the full impact of raindrops and copious runoff. Where these are present gullies develop by:

1 Scouring in the bottom or the sides of the channels.
2 Erosion at the spring head which causes cutting back or headward erosion.
3 Slides or mass movement of soil into the channel from the sides, caused by the lubricating action of seepage, alternate freezing and thawing or undercutting by channel flow.

Gully development stops when any of these factors change markedly, e.g. if the slope angle decreases, say, by the accumulation of fallen debris at the footslope; if vegetation re-establishes itself and so improves soil stability; or if rainfall decreases so markedly as to affect runoff capacity. The processes are all well illustrated in the gully areas of Nigeria, particularly in parts of the Anambra State, the Zaria area of Kaduna State, and the Gombe area of Bauchi State. A small but typical example of gully in the Ibadan area was recently described in the *Nigerian Geographical Journal* (see Faniran and Areola, 1974). The Anambra State examples are obviously the most documented while the gullies around Zaria are also being studied by Dr K. Ologe of Ahmadu Bello University, Zaria. Because they provide examples from contrasting environments, the Anambra State and Zaria examples are here discussed as examples of gully development in humid tropical environments.

The Spectacular Gullies of Nigeria

The Enugu area, or that part of Anambra State which stretches from north of Nsukka to south of Okigwi, has been described as one with an 'unfortunate claim to fame' in having 'some of the most spectacular examples of soil erosion and "badland" topography to be seen in West Africa' (Floyd, 1965, p. 33). This phenomenon has been attributed to both human and physical factors. The main point about the human factor is the high population concentration leading to widespread vegetation depletion which then exposed the deep loose soils. Footpaths, clearings for settlements and market places, as well as excavations are other aspects of man's role in accelerating soil erosion in the Enugu area. Equally important are the physical factors which have been aptly described by Ofomata (1965) among others. He wrote as follows of the physical setting of the area (pp. 49–52):

> Yet it is not the total but rather the nature and intensity of the falls which are important. Generally, the rains come in the form of intensive, violent showers of short duration, especially at the beginning and end of the rainy season. While rain may fall continuously for two, three or more hours ... most of it comes during the first 40 minutes ... the type of rainfall which causes so much damage in a relatively short time, especially in places where the soil is bare or is partially covered by vegetation.... The early rains of February, March and April ... which come just after the end of the dry season are very effective in the process of erosion. The low relative humidity characteristic of the dry season leaves the surface of the soil dry and cracked at various points. These 'fractures' are rapidly exploited by the run-off of storms of the 'early rains' and greatly favour the inception and subsequent evolution of gullies.... Slopes in the Enugu area are, on the whole, relatively steep (particularly in the scarp zone where the gullies are concentrated). The predominant component of the soil is sand ... the 'clay and silt' components vary on the average from less than 4 per cent on the surface to about 30 per cent at a depth of about 8 feet (2·7 m). These finer materials hold the sandy components together but their relatively small proportion emphasizes their inability to prevent the soil from remaining essentially coarse even at depth.

In short, the physical environment is itself right for soil erosion. The 'trigger force' was offered by human interference and vast areas have been rendered useless to agriculture due to extensive gullying action. Ebisemiju (1976) arrived at virtually the same conclusion when, after rigorous statistical analysis, he isolated the soil/vegetation factor as the most important control on landform evolution in the same general area. Vast sums have been expended on checking the erosion and rehabilitating badly eroded areas (see below).

Another area of spectacular gully development in Nigeria is the Zaria area. Zaria lies about 800 kilometres north of Enugu, and has a slightly different climate, vegetation, and geologic setting. The population density is also not as high as in the Enugu area. Yet, the gullies there, although relatively more recent, are just as spectacular as those of the Enugu area.

The important factors of the Zaria gullies seem to be the seasonal rainfall regime, the open (sudan-savanna/woodland-savanna) vegetation and the thick alluvial and largely unconsolidated alluvial fills of prior stream valleys. In other words, what is happening in the Zaria area is the removal of a previously deposited river alluvium by a rejuvenated drainage system. The immediate cause of this accelerated rate of soil erosion is also traceable to human interference in the Zaria region. But while the Enugu gullies are mostly ageing, most of the Zaria ones are still young, fresh, and active.

Some Effects of Soil Erosion by Water

Soil erosion has pronounced effects on the physical processes of infiltration and surface runoff and so on the water-holding capacity of soils; it also affects farming and farming practices, e.g. spray irrigation, drainage, silting, loss of seed, fertilizers, and feed.

SOIL EROSION, INFILTRATION, AND SURFACE RUNOFF

The main effect of soil erosion on infiltration and surface runoff is to decrease the former and increase the latter. Decrease in infiltration is effected mainly by splash erosion which among other things causes muddy surface waters and breaks down soil clods and aggregates. These processes seal up the soil pore space and so reduce infiltration. Reduction in infiltration means more water available for runoff and consequently increased rill and/or gully erosion. Rill and gully development leads to the draining of small surface basins, further increasing the flow velocity.

FARMING, FARMING PRACTICES, AND SOIL EROSION

Soil erosion affects a number of farming practices. First, splash erosion destroys clods and aggregates of soil and causes the puddling of surface materials to form a slowly permeable soil layer on the outer surface. This situation tends to seal up the soil pore space and may even render the surface waterproof. The situation is further aggravated by spray irrigation. It is therefore necessary to know the extent of rainfall impact in order to decide the best method of irrigation.

Soil erosion also has implications on drainage problems. By increasing runoff, larger drainage channels are needed to effect good drainage of swampy areas. Moreover, erosion on the watershed tends to accelerate silt deposition and restriction in channel areas. The silt may also be colonized by vegetation, to further impair channel capacity. Furthermore, accelerated soil erosion results in the silting of dams and reservoirs, the size of the material involved in the silting depending on the location of the dam in relation to the source regions of the sediments. The coarse grains are deposited nearest the source, and the finest farthest away.

Soil erosion affects the topmost layer and so causes seeds and fertilizers which have been spread on the surface to be removed. The extent of such losses depends on the nature of the rainfall. Gentle rains do not cause much splash erosion and so do not remove seed and fertilizer treatments. By contrast, violent outbursts or high-intensity rainfalls are very destructive. Finally, splash erosion and the sealing of pore space of surface soil leads to soil compaction. This make tillage more difficult; it also affects soil aeration and soil quality.

Soil Erosion by Wind

Like water erosion, wind erosion has been active throughout geologic time but it has become much more active and more destructive because of the activities of man, animals, insects, etc., which have removed or considerably depleted the natural vegetation. However, unlike water erosion, wind erosion is most active in arid and semi-arid regions where the land surface is dry and vegetation sparse or absent. Besides, wind erosion may occur wherever soil, vegetation, and climatic conditions are conducive to free wind action, e.g. where:

(a) the soil is loose, dry, and reasonably fine grained;
(b) the soil surface is fairly smooth;
(c) vegetative cover is sparse or absent;
(d) the area involved is sufficiently large;
(e) the wind is sufficiently strong to initiate soil movement.

In any particular area, some of these factors may favour while some may hinder wind erosion. But before we go on to consider these factors, let us look briefly at the process of wind erosion.

THE PROCESS OF SOIL EROSION BY WIND

This process is similar to that of water in that it involves entrainment (removal), transportation, and deposition of soil particles. However, since the soil surface, to be liable to soil erosion, must be loose to start with, the process of detaching soil particles is not important in wind erosion. Instead, three different phases are recognizable, namely, dislodgement or initiation of the movement of soil particles, their transportation, and their deposition.

The initiation of soil particle movement is effected by the turbulence of the wind. The minimum wind velocity required to initiate soil movement is called the threshold velocity, influenced principally by the size of the particle. This value is lowest for grains of 0·1–0·15 mm diameter, which require a velocity of 13–14 km/h at 15 cm above the ground surface. The value increases with either an increase or a decrease in grain size. Thus soil particles of coarse silt to fine sand calibre are most susceptible to wind erosion. The high resistance of finer particles is due partly to cohesion and partly to the fact that such particles are too small to protrude above the laminar or viscous layer of air close to the ground surface. Another factor which affects the threshold velocity is the state of the surface. The threshold velocity for undecomposed crop residue and weeds is higher than for most of the erodible soil grains. This means that such soils are protected until the residue is removed. Similarly, the threshold velocity for bare fields and fields that are unprotected by crop residues from the start is much higher for the first windstorm than for succeeding ones. Once particles begin to move, a 'dynamic threshold velocity' lower than the static or initial one maintains continued movement downwind.

If turbulence initiates the movement of soil particles by wind, wind eddies keep them aloft. As in the lifting, the size of the soil grains involved is very important. It has been found that the rate of sand movement varies directly with the size range of the erosive grains and as the square root of their average diameter. Apart from the soil particle, the gustiness of the wind is also important. Thus, the rate of soil movement is said to depend partially on the density of the air and predominantly on the drag velocity as well as the degree of gustiness of the wind.

Wind is potentially capable of carrying vast quantities of material. The estimated load of sand and dust that may be carried by wind in the atmosphere ranges from 60,000 tons (53,000,000 kg) per cu km of air, depending on the velocity of the wind. At an average wind speed of 48 km/h the carrying capacity of some dust storms for a 24-hour period has been estimated to be more than 1,300 million million kg while a dust storm on 25 March 1895 (in the United States) was estimated to have loaded the atmosphere with 675,000 kg of dust per cu km of air (Stallings, 1957, p. 79). The weight actually involved varies directly with the average diameter of the particles as shown below:

Average diameter (mm)	Weight (g) of soil/cu m of air
0·08	8·40
0·04	23·93
0·007	49·54
0·001	222·53

This will take us to the forms of particle movement by wind or the

SOIL EROSION AND CONSERVATION

forms of wind load which, as in the case of river load, are tractive load or surface creep, saltation load, and suspension load. The biggest particles are dragged along without lifting. Sand and fine gravels are carried in leaps and bounds, i.e. by saltation, while silt and other fine particles are carried aloft in suspension. Measurement of the concentration of wind-borne particles indicated that most of the soil movement in saltation was carried below the height of 10·90 cm, while over 90 per cent of the wind load was transported below the height of 30 cm, and over 50 per cent below the 5 cm level.

The erosive power of wind is provided by the load it carries. Since this increases with distance away from the windward edge of eroding fields, the rate of soil movement also increases in the same direction to a certain limit. Also as soil particles are being moved along, abrasion takes place on impact, the broken particles supplying further tools for wind erosion.

Finally, materials carried either in suspension or by saltation are deposited when the wind velocity decreases. This happens either as a result of obstacles or obstructions on the path of the wind or in more humid conditions at the edge of deserts. Sand dunes and loess are depositional features of wind erosion.

Apart from wind velocity, soil and surface conditions affect the rate of soil erosion by wind. In connection with soil conditions, texture, structure, and the degree of surface compaction or consolidation are important. Since the sand size fractions dominate the process of saltation, it follows that the higher the percentage of sand in a soil the greater its susceptibility to wind erosion. In addition, low silt, clay, and organic matter contents, as obtains in most arid and semi-arid regions, makes clod formation difficult.

By contrast, consolidation of the surface soil particles either as a crust or as clods helps to prevent erosion. If the surface crust is broken into clods which remain on the surface, the degree of erodibility is still low. But if the crust is pulverized sufficiently or is buried, erodibility may increase. Thus, repeated tillage of dry soils usually increases erodibility, especially if it involves pulverization. Action of animal hooves also has a similar effect.

In addition to roughness of the surface contributed by clods or aggregates, wind velocity is affected by ridges and depressions formed by tillage implements or by other causes. Such a surface stops the normal build-up of eroding materials downward and causes wind deposition in the depressions. Similarly, living or dead vegetative matter protects the soil surface from the action of winds. Not only does it reduce wind velocity at the surface by increasing the surface roughness, it also absorbs much of the force exerted by the wind. Soil particles are also trapped. Finally, wind barriers such as shelterbelts, snow-fences, hedges, etc., reduce wind velocity near the ground surface and thus check wind erosion.

DAMAGE FROM WIND EROSION
Damage caused by wind erosion is of numerous types. First, storms over villages, towns, and cities cause untold inconveniences and sometimes serious illness, sometimes even resulting in death, from prolonged dust inhalation. Secondly, fences, ditches, channels, etc., may be blocked or buried and farmsteads or even large settlements may be adversely affected. Thirdly, railways, roads, and other routeways are sometimes blocked by drifting sand or soil, thus increasing maintenance costs. Fourthly, crop damage may result from blowing soil, while the removal of soil to expose tender plant roots or ungerminated seeds may cause crop failure. Whole fields of established crops or pasturage may also be completely covered by drifting soil. Finally, wind erosion, like sheet erosion, can scrape the entire soil cover off an area, as happens in the rock deserts. Indeed, the damage caused by wind erosion may be as serious as, or even more serious than, that caused by water erosion. It is therefore necessary to control both types of soil erosion. The attempts made so far in both these directions are described next.

Soil Conservation Efforts

Water erosion and wind erosion constitute a serious danger to agricultural land. And as observed recently in the United States, loss of soil by wind and water erosion has been severe enough to have lowered the productivity and thus increased the cost of production on over 50 per cent of the agricultural land, while average annual loss by erosion exceeds the most liberal estimates of the amount of soil formed each year by as much as a hundredfold (Stallings, 1957, p. v). Yet the United States is one of the countries where scientific farming with conscious management devices has been known and practised for many decades. The United States is also a young country. The extent of soil loss due to erosion in the old world, especially in those areas practising generally primitive systems of farming, may therefore be imagined. Vast areas in Africa and Asia in particular have been rendered waste by soil erosion. The present-day campaign for the conservation of natural resources is therefore understandable.

The objectives of modern soil conservation are twofold: (1) to reduce to a minimum the accelerated loss of the soil that attends the use of the land for agriculture and other purposes of man; and (2) to attempt to find ways of reclaiming already eroded wasteland. There are three main ways in which these objectives are being pursued. The first is the inculcation of soil management ideas and practices into routine farming. This involves the wise use of the land in order to conserve its natural fertility by growing the right types of crops at the right time and in the correct sequence. Use of fertilizers and manures, mulching, and controlled (minimum) tillage practices are also part of this first approach to land or soil conservation. The second approach concerns the use of certain basic soil erosion control practices, such as contour farming, strip cropping, terracing, etc. In peasant agriculture, shifting cultivation, bush fallowing, and use of animal and other manures are practised for similar reasons but to less effect, while windbreaks and shelterbelts, controlled grazing, etc., are the common practices in the dry regions to check soil erosion by wind. Finally, a number of steps are taken to reclaim areas which have already been badly eroded, e.g. gullies and other badland phenomena. Afforestation, reservation of whole area to prevent encroachment by man and animals, and conscious filling of pits, etc., are the main features of this last approach. Some of these practices apply to both wind and water erosion, but because different factors are involved in each case, and also because they were treated separately in the previous section, they are also considered separately in what follows:

Water Erosion Control Measures
The proper control of soil erosion by water depends on a proper understanding of the causes, factors, and processes of such erosion. As previously stated, soil erosion by water results from the application of energy from two distinct sources—falling raindrops and surface flow. The energy from raindrops is applied from above; therefore remedial measures which intercept and de-energize the raindrop before it strikes the ground are most effective. By contrast, energy from surface flow is applied parallel to the surface. This means designing measures which check the concentration as well as retard the movement of free water over the ground surface. There are measures which serve both ends, and others which serve one or the other purpose. Generally, those that involve increasing the soil cover serve both purposes while practices such as contour farming, building of cross bars, terracing, etc., are intended to check soil erosion by surface flow. These measures are described briefly next, grouped according to the three approaches mentioned above—cropping management, farming practices, and land reclamation practices.

CROPPING MANAGEMENT PRACTICES
The aim of these practices is to control erosion on agricultural land while growing crops. They are therefore simply good (scientific) farming practices, undertaken in full realization of the possible effects of farming the land. The commonest ones, as listed above, are proper land-use practices, cropping systems, use of cover crops, fertilizers and manures, mulching, and proper tillage techniques.

It is important to know the land we are using for agriculture very well. This knowledge includes its capability class and productivity level. This will help us in choosing (a) the land that is suitable for a particular type of cultivation and (b) the best type(s) of crop(s) suited for each piece of land. Few land capability studies have been carried out anywhere in the world, while those that have been done, e.g. for Nigeria (see FAO, 1966), are so generalized that they are not very useful at the level of an individual farm. Nevertheless, because they provide some information about the class of the land and show whether or not the land is suited for cultivation at all, they can be very useful in planning soil conservation projects.

Another important feature of good agriculture to conserve the soil is the cropping system. The system of crop or farm plot rotation adopted is very important. This may be a three-, four-, or five-year system, involving different types and combinations of crops. Crop rotation is a significant feature of agriculture in the Western world in particular, while modifications of it—e.g. the farm rotation, or the bush fallow system—occur in the less developed world. This is because of the known benefits, which include:

(a) reduction in erosion resulting from the high degree of protection provided especially by the cover or sod crops;
(b) improved soil structure produced by the root system of the sod crops and the relatively large amounts of organic matter that are returned to the soil;
(c) increase in soil nitrogen resulting from nitrogen fixation by the legumes in the system.

There are cases where crop rotation is not the most economic use of land, because growing a single crop year after year ensures maximum productivity. Examples are the Corn Belt and the Wheat Belt of North America among others. It is important to note that continuous cropping is only feasible where the soil is exceptionally fertile and also where a number of measures are taken to maintain this level of fertility. Where the land is prone to erosion, e.g. as a result of slope, crop rotation is practised. Fertilizers, manures, mulching, etc., are also heavily applied. Otherwise the land will degenerate, as happened in the Cotton Belt of the United States.

Finally, the machinery or equipment used in farming is an important aspect of good agriculture. Different types are now available on the market and it is essential that we use only those which will do the minimum of damage to our soils, in order to minimize erosion.

FARMING METHODS AND EROSION CONTROL
These methods are those which are used to support normal cropping practices to minimize soil loss on agricultural land. The commonest ones are contour farming, strip cropping, terracing, diversions, and other

structures which check runoff on farm plots and so decrease soil erosion, usually in steeply sloping areas.

Contour farming is the practice of planting rows or operating farm equipment across the slope. The method helps to conserve soil and water as the rows of ridges or crops, or both, act to check water flow. The effectiveness of the method is shown by the following statistics from field tests in the United States: soil losses as well as water losses (on 4–6 per cent slopes) were reduced by up to 50 per cent, while crops yields were increased as follows: maize 7·3 bushels per acre; soybeans, 2·5 bushels; oats, 5 bushels (FAO, 1965, p. 90).

Strip cropping is the practice of planting alternate strips with close-growing meadow, row, or grain crops following the contour across a slope. The system is used on slopes that are too steep to terrace. Like contouring, strip cropping slows down runoff water flowing through the close-growing strip and so increases the infiltration rate, which further reduces total runoff. The crops also intercept raindrops and so reduce both splash and surface flow erosion.

Terrace agriculture involves building earth embankments or a combined channel and embankment across the slope, usually at fixed intervals. It may also involve creating flat or near-flat surfaces along very steep slopes. The latter is called bench terrace and is a much more tedious task which is undertaken where embankments and channels cannot stop runoff on very steep slopes. Normal cultivation is also impossible along such slopes which are difficult to climb. The best examples of these terraces known to the authors are in the Aku area of Anambra State, between Enugu and Nsukka, and those around the village of Mamu near Awgu. Here, the flat hilltops are capped by hard duricrusts, while the valley bottoms are covered by palm bush. The hillslopes have deep red earth. Terrace farming is also widely used on the Jos Plateau, Nigeria, as well as in many mountainous areas of the world. Israel, Kenya, and the Andes region of South America provide good and even better examples outside Nigeria.

Terraces, whatever their type, control erosion by (a) reducing slope length as well as slope steepness (bench terrace), and (b) slowing down runoff. Runoff may also be conducted across the slope in definite channels where it flows usually at non-erosive velocity. This latter system is a sort of diversion technique, by which specially designed channels are constructed to conduct water to a safe outlet. This is particularly important in gully areas where such channels are made to divert water away from areas of active gullying activity. Failure to take proper steps has resulted in the stimulation of gully activities as recently reported by Nir and Klein (1974). This happened in the Nahal Shiqma area of Israel where terracing has accelerated the rate of gully development.

Finally, a number of structures are in use which help to check soil erosion and conserve water. These vary from mounds or cross ridges in furrows to weirs and spillways.

RECLAMATION OF BADLY ERODED LANDS
There are vast areas where the above control measures are either not known or are ineffective. In such areas, thousands of acres of farmland have been rendered useless by accelerated soil erosion. The most spectacular ones are the badly gullied areas of the Anambra State of Nigeria. Badland topography is also widespread in other parts of Africa and the world, while sheet erosion has scraped vast areas of the best part of their soil cover. In the past, and also in countries where population is scanty, such areas used to be abandoned for other virgin areas. However, population has reached critical levels in many countries, which cannot now afford such a 'luxury' as shifting cultivation. Man has also become less nomadic, living in established towns and/or on individual properties. Waste lands therefore have to be reclaimed.

Fortunately, the world at large has learnt a lot from the US experience in parts of the Cotton Belt in the States of Georgia and South Carolina, among other places. Here large areas have been successfully reclaimed, using certain standard techniques including re-afforestation, drain diversions and the construction of new drains, and the prohibition of human activities in badly eroded areas. A very important aspect of this is the degree of co-operation between the local people and the reclaiming agent, which in most cases is the government or its functionary. That the most successful projects are often undertaken by or with the co-operation of the local people can be illustrated by the Nigerian experience.

SOIL CONSERVATION IN THE ENUGU AREA OF NIGERIA
The Enugu area has been shown to be one of the worst-affected areas as far as gully development is concerned. It therefore offers a good example of the successes and failures of anti-erosion campaigns in a developing country. The story relating to this topic has been told several times (see Sykes, 1940; Floyd, 1965) and only the most important points need be mentioned here, as follows:

(1) Little or nothing has so far been done by the local people, who appear helpless. As Floyd (1965, p. 39) observes, 'Once the erosion gullies had appeared, communities were powerless to check their growth, if indeed any desire arose or efforts were made in the early stages to arrest their development.'

(2) There seems to be no conscious attempt at an overall conservation scheme or programme covering the entire region. The major efforts have been localized and restricted to a few areas. Examples of *ad hoc*, unco-

ordinated projects are the establishment of the Udi Forest Reserve in 1918, followed in 1923 by a project designed to check a particular gully which was observed to be threatening the Government Station at Udi. To do this, it was recommended that (a) the drain be diverted and new drains constructed, (b) certain footpaths be closed and others drained, and (c) grazing be prohibited on hillslopes. Again in 1928, a number of measures were taken, including (a) the construction of ditches and ridges at 25-cm intervals between the ravines and (b) the introduction of exotic plants to stabilize both the ridges and the ravines. The ridges were planted with trees—e.g. the cashew nut tree—and the ravines with bahama grass. The next move was in 1937 when an area of about 30 hectares was involved. The project consisted of (a) wave-bedding round the head and sides of the ravine (the same ravine mentioned above as threatening the Udi administrative office building), (b) construction of a number of dams in the ravine to check excessive runoff—a total of 131 dams was recommended—and (c) the stablilization of wave beds by direct sowing of seeds.

(3) There is a wide gap between the government (the innovators) and the people. This is evidenced by the unwillingness of the local people to adopt the various innovations. For example, attempts to enforce contour farming and the use of ridges in place of the traditional heaps were generally viewed with suspicion and sometimes with hostility. Similarly, new farming methods such as crop rotation, planting of cover crops, mulching, use of fertilizers and manures, etc., appear to have been rejected by most of the people.

(4) A number of problems militate against the conservation attempts. One such problem is the general apathy and lack of co-operation of the local people, caused among other things by (a) problems created by political agitators, especially the so-called socialists, who preach that it is the government and not the people that should check the gullies, (b) ignorance and illiteracy on the part of the farmers, and (c) the general conservatism of the people. Another problem is that of excessive population pressure on land, this area having one of the highest population density figures in the tropical world.

(5) Finally, there is the problem created first by party politics and then by the civil war, which caused so much unrest and dislocation of human activities in the area. These have definitely had adverse effects on the government programmes.

Clearly the situation in Nigeria contrasts very markedly with that in the Piedmont region of the United States where the problem of accelerated soil erosion has been successfully tackled by a comprehensive programme which involved on the one hand the government (federal, states, and local) and on the other the people. What is needed in the Nigerian case, as in other similar cases in developing countries, is willingness on the part of the government to accept its responsibility to educate and help the people to fight the teething problems, not only of rehabilitating existing badly-eroded areas but also of preventing repetition in other areas of the mistakes which created such gullies. There is a very urgent need for a countrywide campaign and programme for the conservation and wise use of the natural resources of developing countries, who should learn from the mistakes of the developed ones.

Control of Wind Erosion
Like water erosion, the best cure for wind erosion is to fight the cause or causes. Since the most important of these causes is obviously lack of vegetation cover, the best means of preventing wind erosion, as also of reclaiming wind-eroded land, is to provide vegetal cover for the land. Knowledge of the process of wind erosion is also an important consideration in the attempt to check it. It is on the basis of this understanding that the following methods of control have been recommended by the FAO (1960):

(a) the production at the soil surface of aggregates or clods which are large enough to resist wind action;
(b) increasing the roughness factor of the land surface to reduce wind speed and trap drifting sand;
(c) establishment of barriers or trap strips at intervals, also to reduce wind speed and soil avalanching;
(d) establishment and maintenance of vegetation or other types of cover (crop and other vegetative residues, for example) to protect the soil surface.

These principles of control have general application but the relative importance of each varies from place to place, depending on climate, soil, and, perhaps most importantly, land-use conditions. The common land uses are cultivated (cropped) lands and grazing land, while large areas cannot be used at all. We shall discuss the various methods under these different land-use conditions.

CONTROL OF WIND EROSION ON CULTIVATED LANDS
The growing of crops provides vegetative cover for the soil during the time of their growth. Where row crops are involved, the best system is to orient the rows perpendicular to the prevailing wind. Additional protection may also be needed, e.g. in the form of plant residue between the rows. Also after harvesting the stumps should be left at sufficient height to check the wind speed.

Strip cropping is another recommended system. Here, narrow strips of erosion-susceptible and erosion-resistant crops are alternated, with their lengths running perpendicular to the direction of the prevailing wind. The systems in use in the United States include wheat/fallow and wheat/sorghum/fallow, etc. Any of these systems will help to trap saltating soil particles and so control avalanching. Since soils vary in their textural characteristics and so in their erodibility, different widths are recommended for different soil classes (Table 10.2). The widths are based on wind velocity of 64·4 km (40 m/h) at a height of about 15 m, blowing at right angles to the strips and stubble 30·5 cm high upwind from the erodible strips (FAO, 1960, p. 19).

TABLE 10.2
Relationship Between Strip Width and Soil Textural Class

Soil class	Width of strip (m)
Sand	6·1
Loamy sand	7·9
Sandy loam	30·2
Loam	75·5
Silt-loam	85·6
Clay-loam	105·4
Silty-clay-loam	130·8
Clay	25·0

Strip cropping has a number of disadvantages, e.g. weeds growing at the edge of the strips, insect infestation, soil accumulation at strip edges, added costs, grazing difficulties, and so on; but most of these can be avoided with careful management. Although dry regions are not ideally suited to it, crop rotation of some sort, by keeping the land fertile, also helps to control wind erosion. Other aspects of good agriculture which helps to control and perhaps prevent wind erosion are the establishment of windbreaks and shelterbelts, the proper management of crop residues, and the careful use of tilling instruments.

CONTROL OF WIND EROSION ON GRAZING LANDS
The main cause of wind erosion on grazing lands is overgrazing; the importance of controlled grazing, therefore, cannot be overemphasized. The animals definitely have to be provided with alternative feeding stuff during periods of poor vegetative growth. Another cause of wind erosion is excessive trampling of the soil caused by over-concentration of large numbers of animals on the land, which not only destroys the vegetation

but also breaks loose the surface soil and so renders it susceptible to wind erosion. The solution to these problems may be found by:

(a) supplying adequate watering sites, so that the animals can be moved constantly;
(b) fencing or otherwise excluding animals from very erosive parts of the pasture;
(c) avoiding placement of gates or lanes on erosive sites or moving them when an erosion problem starts;
(d) providing wind barriers to protect permanent lanes, water sites, etc.

The most important factor is therefore vigilance so that corrective measures are applied as soon as erosion starts.

APPENDIX

LABORATORY TECHNIQUES OF SOIL ANALYSIS

Few university geography departments in the world have a laboratory which is adequately equipped for real soil analysis. This is due to a large extent to the fact that soil geographers have not made up their minds about the desirability of laboratory soil tests. Those who are convinced of the need for them are not quite sure how far a geographer should concern himself with what is generally regarded as the domain of the pure soil scientist. To many people, soil geography involves no more than the mapping and classification of soils, while many standard texts on soils, written by soil scientists and agronomists, characteristically have a section on 'soil geography' which is nothing but a list of place-names where particular soil types or categories of soils are found.

But there is more in soil geography than this. Geographers should be interested in the entire field of *pedology*, embracing the factors and processes of soil formation, the description and analysis of soil morphological characteristics, and the classification and systematic mapping of soils. They are not concerned with pure soil science *per se* which is a field dominated by chemists who regard the soil as a vast chemical laboratory, nor with edaphology, the standpoint of the agronomist. None the less, geographers are coming to realize that in order to understand soil formation, the role of individual factors and processes, and spatial differences between soils, they need to undertake the routine laboratory tests of soils and carry out the more fundamental micro-morphological and mineralogical analyses. Some of this work will have to be done at the postgraduate and research levels. But, for undergraduate teaching, a soil geography laboratory should be equipped to perform experiments on some of the basic physical and chemical soil properties including particle-size analysis, moisture content, Atterberg limits and soil aggregate stability, bulk density, organic matter content, soil pH, calcium carbonate content, soil nitrogen, cation exchange capacity, and the exchangeable metallic cations.

The laboratory analysis of soils can be slow and intricate. Much of the equipment and apparatus is expensive to buy and costly to run. It is no wonder, then, that geography departments are often reluctant to establish soil laboratories when the analysis of soils can be conveniently left in the hands of the agronomists and the soil scientists. But the cost of running a laboratory need not be insurmountable, especially if the arrangement is such that there is a unified physical geography laboratory taking in not only soils, but also geomorphology, hydrology, and biogeography.

Laboratory Equipment and Organization

Apart from the apparatus and chemical reagents for carrying out the different types of analysis, the laboratory should have certain basic facilities. In the first place, it should have a storeroom or shelves and lockers where soil samples can be kept safe from contamination. This is important where the laboratory is being used for other purposes, e.g. plant analysis. Chemical reagents, glassware, and other laboratory equipment are kept on shelves in the laboratory itself or in a separate room. Also, if possible, the laboratory should have another adjoining room where all the 'dirty' work like grinding, sieving, and preparing soil samples for analysis is done. The laboratory itself should be spacious enough to take long tables and platforms where experiments can be laid out and on which soil samples can be spread for drying and examination.

The laboratory should have an adequate supply of running (tap) water, gas, and electricity. A water distiller is essential as distilled water is used in nearly all experiments.

A soil laboratory requires at least two types of weighing balances: (1) the top-load balance capable of weighing bulk soil samples and (2) the more accurate, automatic digital balance. The latter type of balance is used in weighing fairly small quantities of soil for experiments where such precision is vital to the results. It is also used for weighing the ingredients for preparing the chemical reagents.

Other essential basic laboratory equipment includes: a centrifuge; a water bath that is electrically heated, thermostatically controlled, and fitted with an electrically driven water stirrer; a hot plate; an oven; a desiccator with blue silica-gel granules or fresh calcium chloride; a magnetic stirrer; a reciprocal shaker; and a mechanical sieve shaker.

Methods of Soil Analysis

There is usually, for every soil property under investigation, more than one method of analysis. The methods described below have been chosen because they are comparatively easy to understand and are not time-consuming. The equipment outlay is neither too expensive nor unduly complicated. The method adopted for a particular soil test may not necessarily be the best method for it but it is sufficiently accurate for our purpose.

1 *Moisture Content*

The determination of soil moisture in the laboratory presupposes that the sample has been well preserved, i.e. that there has been no gain or loss in moisture content since it was taken from the field. Two experimental methods of determining soil moisture are described: (a) the oven-drying method and (b) the electrical conductivity method.

(a) Oven-drying method
Apparatus:
(i) An oven at a controlled temperature of 105–110°C
(ii) Heat-resistant glass weighing bottles with glass stoppers (or crucibles of suitable size)
(iii) A desiccator (with blue silica-gel granules)

Procedure:
(i) Take a glass weighing bottle and weigh it with the lid on to obtain Wc.
(ii) Put a certain quantity of soil, say, between 20 and 100 g, in the bottle; stopper the bottle.
(iii) Weigh the bottle and its content to obtain Ww.
(iv) Remove the stopper, place the bottle in an oven and dry to constant weight at a temperature of 105–110°C.

The drying period will vary with the type of soil but is usually 16–24 hours. Peat and peaty soils may require longer periods since these are usually dried at a lower temperature (e.g. 60°C) to avoid oxidation of the organic matter.
(v) After drying, allow the bottle to cool in a desiccator.
(vi) Stopper the bottle and then weigh it together with its contents to obtain Wd.
(vii) Calculate the percentage moisture content (M) by the formula

$$M = \frac{Ww - Wd}{Wd - Wc} \times 100.$$

(b) Electrical conductivity method
Here, the moisture content is determined by measuring the electrical conductivity of a salt solution (e.g. CaCl) in equilibrium with the soil. This method is based on the ability of electrolytes, usually present in the soil solution, to conduct electric currents. The higher the concentration of electrolytes, the greater the conductivity. The lower the water content, the higher the concentration of electrolytes. The reciprocal of conductivity is resistance which increases with the degree of dilution. This electrical resistance when measured accurately can provide an estimate of soil moisture as shown below (cf. gypsum block method described in Chapter 6).

Apparatus and Reagents:
(i) Electrical conductivity measuring instrument in Ohm units—the Wheatstone bridge
(ii) Test tubes
(iii) 5×10^{-1}M calcium chloride solution
(iv) Water bath at a constant temperature of 25°C

Procedure:
(i) Weigh out, say, six samples of the oven-dried experimental soil, 10.0 ± 0.1 g each, into clean, dry test tubes.
(ii) Add 1·0, 2·0, 4·0, 6·0, 8·0, and 10·0 ml distilled water respectively to the six tubes. Find the weight of the wetted soil to obtain percentage moisture content.
(iii) Pipette 10 ml of 5×10^{-1}M calcium chloride solution into each tube. Shake very well.
(iv) Place the tubes in a water bath at 25°C.
(v) When the temperature equilibrium has been achieved (i.e. when the temperature of the solutions in the tubes has also risen to 25°C), and the particles have settled to the bottom of each tube, measure the electrical resistance of the solutions starting with the most dilute one. The readings are in Ohm units.
(vi) Plot, on a graph sheet, the resistance readings against the percentage water content of the soil.
(vii) From this curve read off the moisture content of the fresh sample of the soil. This is done by first measuring the electrical resistance of the fresh soil sample at the same temperature (25°C) as that of the standard solutions.

2 Atterberg Limits

These apply to the fine-textured soils whose physical state is affected by the water content. Thus a soil may exist in four physical states: *liquid, plastic, semi-solid, solid*. The boundaries between these states, known, respectively, as the liquid, the plastic, and the shrinkage limits, were defined empirically in 1911 by A. Atterberg, and so are known collectively as the 'Atterberg limits'. These limits are important in determining ground conditions: its stability, firmness, and/or trafficability, especially for engineering purposes. Hence they are also referred to as the 'consistency limits'. The limits are defined by the amounts of water required in the soil to produce specified degrees of consistency. The determinations of the liquid and plastic limits only are described, being the most common of the tests usually carried out in the laboratory.

(a) *Liquid limit test*
The liquid limit is the point at which the soil becomes semi-fluid. It is the moisture content at that stage at which a groove cut in a moist soil

sample held in a special brass cup is closed after twenty-five jarring taps or blows on a rubber plate. The jarring of the cup is achieved by using an apparatus which will lift and drop the cup precisely 1 cm.

Apparatus:
(i) BS sieve no. 36 (mesh 0·422 mm)
(ii) A glass plate about 9·5 mm thick and approximately 0·6 m square
(iii) A small spatula
(iv) A liquid limit apparatus: a brass cup and carriage mounted on a rubber base 18 cm × 13 cm × 5 cm (approximately)
(v) A standard grooving tool (Casagrande type) with 'gauge' handle
(vi) Moisture content test apparatus (as above)

Procedure:
(i) Take a representative sample (about 120 g) of air-dried soil sieved through BS no. 36. Place it on the glass plate and with the spatula mix it thoroughly with distilled water to make a thick paste. Leave to mature for two hours.

The amount of water that will be required to make the air-dried soil into a thick paste will vary with the type of soil. It is better to proceed cautiously by adding the water in small quantities.

(ii) Put a small quantity of the soil paste in the brass cup and level off the top with the spatula, so that it is parallel to the rubber base. The maximum depth of soil in the cup should be 1 cm. Some of the paste from the cup is set aside for moisture content determination.

(iii) Using the grooving tool divide the paste in the cup along the cup diameter. This will leave a V-shaped gap, 2 mm wide at the bottom, 11 mm at the top, and 8 mm deep.

If the groove fills immediately, it means there is too much water in the paste, in which case the whole process will have to be repeated.

(iv) Check the liquid limit apparatus to make sure the cup drops exactly 1 cm when the handle is turned. Then, by turning the handle at the rate of two rotations per second, the cup is lifted and dropped. Continue to do this until the two sections of the paste merge for a length of 1·3 cm along the bottom of the groove. Record the number of blows at which this occurs.

A figure of over fifty blows is unacceptable and the experiment should be repeated putting in more water.

(v) Repeat the experiment twice more, taking a different sample from the paste each time without adding more water. The results should be nearly the same. Find the average number of blows.

(vi) Take a sample each from the portions of paste set aside for moisture content determination and put them all in a single glass

weighing bottle. Then take a sample from either side of the groove and place it in the same bottle, which will now contain five samples. Find the moisture content.
(vii) Repeat processes (i) to (vi) four times, using the same soil but adding distilled water at the beginning of each stage.
(viii) Plot the average number of taps against the corresponding moisture content on a semi-logarithmic graph sheet. A straight line, the flow curve, is drawn through the points. The moisture content corresponding to the twenty-five-blow ordinate can then be read off the flow curve. This gives the liquid limit of the soil.

(b) *Plastic limit test*
The plastic limit (PL) of a soil is the moisture content, expressed as a percentage of the dry weight, at which the soil begins to crumble on being rolled into threads 3 mm in diameter.

Apparatus:
The same as for the liquid limit test except that there is no need for the liquid limit apparatus and the grooving tool.

Procedure:
(i) Take 15 g of soil sieved through BS no. 36 and mix it thoroughly on the glass plate with sufficient distilled water to make it plastic enough to be shaped into a ball.
(ii) Roll the ball into a thread on the glass plate with the palm of the hand. When the thread has been rolled to 3 mm diameter, the specimen is kneaded together and rolled out again. Continue to do this until the thread crumbles.
(iii) Collect the crumbled threads in a glass weighing bottle and determine the moisture content.
(iv) Repeat tests (i) to (iii) twice more. The average of the three moisture contents is taken as the plastic limit of the soil.

Note: Plasticity Index $= LL - PL$

Liquidity Index $= \dfrac{M - PL}{PI}$

where M = moisture content (%), PL = plastic limit, PI = plasticity index.

3 *Bulk Density* (density–can method)
Bulk density is the ratio of the weight of soil to its volume expressed in grammes per cubic centimetre. The bulk density of the soil influences its degree of compaction, bearing capacity, and the stability of the ground. The 'weight' of a bulk soil sample is easily measured. It is the measurement

of the 'volume' that poses the real problem. A number of methods have been employed in measuring 'volume' but the simplest and easiest to understand is the 'density-can method'.

Apparatus:
(i) A density can: a cylindrical metal container filled with a siphon tube, the outlet of which is fitted with a rubber tube and a spring clip
(ii) A graduated 2,000-ml glass cylinder.
(iii) Paraffin wax, of known specific gravity

Procedure:
(i) Get a solid block sample of soil and trim it into a more or less regular shape avoiding re-entrant angles.
(ii) A graduated 2,000-ml glass cylinder
(iii) Put on with a brush a thin coat of paraffin wax. Allow to dry and then apply a second coating of paraffin wax. Weigh the coated sample to obtain Wx. Hence, obtain weight of paraffin wax (Wp), which is $Wx - Ws$.
(iv) Select a suitable density can (15 cm or 30 cm). Fill it with water to above siphon outlet. Release the clip on the rubber tube outlet for the excess water to run off. Replace the clip.
(v) Place the coated sample in the can, making sure it is totally immersed. The water displaced by the sample is run into the measuring cylinder and the volume (Vw) recorded.
(vi) Remove the coated sample from the can and wipe it dry on the outside. Then remove the paraffin wax coatings. Find the moisture content of the block sample (using a representative sample).

Calculations:

$$Vs = Vw - \frac{Wp}{Gp} \text{ ml}$$

where Vs = volume of soil and Gp = density of paraffin wax (usually 0·908 approximately).

$$r = \frac{62·425}{Vs} Ws \text{ lb per cu ft}$$

where r = wet bulk density.

$$rd = \frac{100 Dw}{100 + m} \text{ lb per cu ft}$$

where rd = corresponding dry bulk density, Dw = dry weight of soil, m = moisture content %.

4 Water-Stable Aggregates

An aggregate is a group of two or more primary particles which cohere to each other more strongly than to surrounding particles. The interest in soil aggregates lies mainly in their size distribution, which is a major determinant of their susceptibility to movement (erosion) by wind and water. They also determine the size of the pore spaces, soil aeration, and drainage in ploughed soils.

The disintegration of the soil mass into aggregates in the field is usually due to the disruptive force of ploughing and the impact of falling raindrops. In determining the size distribution and the stability of aggregates in the laboratory an attempt is made to simulate these natural disruptive forces. Size distribution of aggregates varies with the soil moisture condition, whether dry or wet. Cultivated soils also differ from undisturbed ones in size distribution. The choice of moisture condition under which the experiment is carried out depends on the purpose of the analysis. 'Dry sieving' is recommended when the aim is to determine the susceptibility of the soil to movement (erosion) by wind, while 'wet sieving' yields results relevant to the behaviour of the soil mass when wet or when cultivated.

Size distribution of aggregates by wet sieving

This method involves first wetting the soil sample and then separating the aggregates into various sizes by sieving the soil through a nest of sieves under water. The wetting of the soil can be done in two ways: (1) direct immersion of the dry soil in water at normal atmospheric air pressure, or (2) wetting under a vacuum. The first method causes great disruption of large aggregates into smaller ones. But the method is useful in that this type of wetting is similar to what happens when the soil is wetted by irrigation. The method could also be applied to surface soils. But for the layers below the immediate surface the alternative method is preferred. There is less disruption of the aggregates and the results give a better insight into the geometry of soil particles and voids. The latter method has been adopted in the experiment which follows.

Apparatus:
- (i) Nests of sieves 12·7 cm in diameter and 5·1 cm high with mesh nos 4, 9, 16, and 25; and one $2\frac{1}{2}$-mesh screen (8 mm)
- (ii) If available, a Yoder-type sieving machine which raises and lowers the nests 4 cm through water approximately thirty times per minute.
- (iii) Vacuum desiccators and small barometers
- (iv) Mechanical stirrers
- (v) 75-mm watchglasses

Procedure:

(i) Take the moist soil sample and sieve through an 8-mm mesh, pulling apart clods larger than 8 mm until the subunits are small enough to go through the sieve. Try to avoid breaking clods into units less than 4·76 mm.

(ii) Air-dry the sample at room temperature.

(iii) Weigh out three representative subsamples of the air-dried soil of 25 g each. Oven-dry and weigh one of the samples to determine moisture content. Assume that the other two samples contain the same amount of water as the dry soil.

(iv) Place the two samples not oven-dried in 75 mm watchglasses on the perforated ceramic plate in the vacuum desiccator. Put 5–10 ml of water in the bottom of the desiccator and a small barometer on the ceramic plate.

(v) Prepare a vacuum in the desiccator by turning on the water pump. When the barometer reading has fallen to zero pressure (that is, no mercury remains in the central tube), close the clamp on the rubber tubing.

(vi) Remove the tap end of the tube and immerse it in a bottle containing 'de-aerated' distilled water. The water is allowed to enter the desiccator until it comes over the edge of the watchglasses and wets the samples.

(vii) Remove the watchglasses containing the samples from the desiccator and transfer them into separate nests of sieves which are immersed in distilled water. Then carefully slide the watchglasses away.

(viii) After about ten minutes sieve manually for ten minutes (or if available use a Yoder-type sieving machine).

(ix) Remove the sieves from the water. Oven-dry the samples in each sieve and weigh to obtain W.

Part of the material in each sieve is usually too large to pass through the mesh. Hence it is necessary to determine the amount of sand in each sieve. This is done by immersing the sieve in calgon solution to deflocculate and disperse the soil. After all the fine soil particles have been washed away the sand particles left in the sieve are dried in the oven and weighed to obtain Ws.

(x) Find the weight of aggregates in each sieve by subtracting the weight of the sand from the weight of the oven-dry sample $(W - Ws)$.

(xi) Calculate the quantity of soil material smaller than the finest sieve in the nest by subtracting the sum of the oven-dry weights of material retained in each sieve from the oven-dry weight of the original sample.

Calculations:
Multiple-sieve techniques like the one described above give values of the amount of aggregates in each of several size-grades. These individual values are useful in that it is then possible to assess their relative importance; a soil with a greater proportion of large aggregates is better for agricultural purposes than one with a larger proportion of small aggregates. However, there is a need to express aggregate size distribution of a soil by a single parameter for comparative purposes. One such parameter is the 'mean weight-diameter' (MWD). It is equal to the sum of the products of the mean diameter x_i of each size-grade and the proportion of the total sample weight W_i of each size-grade, i.e.

$$MWD = \sum_{i=1}^{n} \bar{x}_i W_i.$$

To calculate it, divide the weight of aggregates in each of the five size-classes above by the weight of the oven-dry sample minus the weight of the sand. This gives values for W_i. Find the mean weight-diameter (MWD) using the equation.

5 Particle-Size Analysis

Particle-size analysis is the quantitative determination of the relative proportions of sand, silt, and clay in the soil. In engineering, particle-size analysis is usually divided into two main categories: (1) mechanical analysis (sieving) of loose materials like sand and gravel, and (2) sedimentation (pipette sampling and hydrometer methods) tests applied to the more cohesive clays and silt. Since there is no soil which is purely sand or purely clay, the two methods are usually combined in one operation to obtain a complete particle-size analysis of a soil sample.

Before any particle-size analysis can be done the soil sample has to undergo a certain amount of pre-treatment to remove organic matter, calcium carbonate, and other soluble salts which may cause flocculation and hinder the separation of the fine soil particles into individual grains. The pre-treatment applied to soils varies with the type of soil, whether it is loose sandy material or whether it is the more cohesive loamy and clayey soils. Many sandy soils need no pre-treatment as they do not contain much organic matter nor do they contain soluble salts.

PRE-TREATMENT OF SAMPLES FOR PARTICLE-SIZE ANALYSIS
(a) *Removal of carbonates*
Carbonates are removed by washing the soil with acid, such as dilute hydrochloric acid (HCl) or acetic acid (HAC). The latter is more commonly used not only because it is effective in removing the carbonates

but also because it fosters the oxidation process (of organic matter) that follows. Manganese dioxide (MnO_2) is known to be able effectively to reduce the ability of hydrogen peroxide (H_2O_2) in decomposing soil organic matter. Acetic acid, however, reduces the manganese dioxide to the manganous form and so prevents it from interfering with the oxidation process, at least in non-calcareous soils. Also, pre-treatment of the soil with acetic acid makes the possibility of the exfoliation of weathered mica by hydrogen peroxide less likely.

How the leaching is done depends on how calcareous the soil is. Some highly calcareous soils need to be soaked in acid for a period of twenty-four hours or more to remove all the carbonates. For the less calcareous soils the operation is much shorter. The soil sample is transferred into a centrifuge bottle with the required amount of acid added. The mixture is then shaken vigorously, preferably on a reciprocal shaker, to dislodge the carbonates. The solution is then centrifuged to allow the soil particles to settle.

(b) *Oxidation of organic matter*
The commonest method of removing soil organic matter is to heat the soil with hydrogen peroxide (H_2O_2) on a hot plate. Most organic particles in the soil are of the same size-range as the minute clay particles and, if not completely oxidized from the soil, could lead to an overestimation of the clay fraction in the soil. Unfortunately hydrogen peroxide does not effectively remove all types of organic matter in the soil. It is capable of decomposing colloidal or humidified organic matter but not the fibrous (cellulosic) residues.

(c) *Particle fractionation or dispersion*
The next stage before actual particle-size analysis can begin is to break the soil down into individual grains. This is known as dispersion or particle fractionation. There are many chemicals which can be used as dispersing agents but the one most commonly used is sodium hexa-meta-phosphate (or calgon). The strength of the calgon solution to be used depends on the type of soil. Many soils are easily dispersed with 0·1 per cent calgon solution but some very clayey and highly calcareous soils may require a 5 per cent solution.

PARTICLE-SIZE ANALYSIS
This is done by the sedimentation method, two types of which are known—the pipette and the hydrometer methods.

(a) *The pipette method*
The pipette sampling method is the most accurate, albeit a very laborious

and time-consuming, method of particle-size analysis. It involves (1) the pipette sampling of the finer soil fractions, clay and silt, at a standard depth in a 1,000-ml cylinder and at specified times according to the temperature of the suspension, and (2) sieving mechanically to separate the different size-grades of sand particles.

Apparatus and Reagents:
(a) *for pre-treatment*
(i)   1,000-ml conical beaker
(ii) 200-ml centrifuge bottle
(iii) N/2 acetic acid buffered at pH 2·0 with sodium acetate
(iv) 20 vol. H_2O_2
(v)   Hot plate; reciprocal shaker; centrifuge

(b) *for sedimentation*
(i)   1,000-ml graduated cylinder
(ii) An Andreasan pipette, capacity 10 ml, 20 ml, or 25 ml, although an ordinary pipette will do as well
(iii) 400-ml conical beaker calibrated at 10 cm
(iv) Glass weighing bottles or evaporating dishes
(v)   Oven
(vi) Stop clock

(c) *for sieving*
(i)   A nest of sieves for example with mesh 2·0 mm, 1·0 mm, 0·5 mm, and 0·2 mm in diameter
(ii) Mechanical sieve shaker

Procedure:
(i)   Weigh out 20 g of 2-mm soil sample. Too great a weight of soil should not be used in case the suspension becomes too concentrated; a maximum weight of 30 g should do.
(ii) If the soil is particularly calcareous, soak overnight (or for as long as is necessary) in N/2 acetic acid to dissolve all the carbonates. Otherwise, transfer the sample into a 200-ml centrifuge bottle; add N/2 acetic acid and shake vigorously on a reciprocal shaker to dissolve the carbonates.
(iii) Centrifuge the suspension to allow the soil particles to settle. Throw away the supernatant liquid with dissolved carbonates.
(iv) Add distilled water to the sample in the centrifuge bottle and shake to dispersal point on a reciprocal shaker to wash the soil clean of the acid and the dissolved soluble salts. Centrifuge again.
(v)   Transfer the sample into a 1,000-ml beaker and add 20 vol. H_2O_2. Place the beaker on a hot plate.
      The amount of H_2O_2 used will depend on the amount of organic matter in the soil. The reaction of the H_2O_2 on organic

APPENDIX

matter causes frothing. Continue to add H_2O_2 until the frothing has ceased.

(vi) When the H_2O_2 has completely evaporated, transfer the dry sample into a 200-ml centrifuge bottle, add distilled water, shake on a reciprocal shaker to wash away the H_2O_2 and the oxidized organic matter. Centrifuge.

(vii) Add 0·1 per cent calgon, pH 8·0. Shake overnight on a reciprocal shaker to disperse the soil particles.

(viii) Transfer the solution into a 1,000-ml graduated measuring cylinder and make up to mark with distilled water. Take the temperature of the suspension. This will determine the settling times for the separation of the fractions—see Table A.1.

To avoid temperature fluctuations during the experiment some people stand the cylinder in a water bath at a specified constant temperature, say, 25°C.

(ix) Stir the suspension vigorously. Allow to settle for 4 minutes 15 seconds, at temperature 25°C. Lower the pipette into the solution to a depth of 10 cm and withdraw a 25-ml sample. Transfer the sample into a weighed evaporating dish and oven-dry at a constant temperature of 105°C. After drying, weigh the soil and the dish. The difference in weight (wt soil + dish − wt dish) is $\frac{1}{40}$ (25-ml sample from 1,000-ml suspension) of the weight of coarse and fine silt and clay in the original sample of soil.

(x) Immediately after taking the sample at 4 minutes 15 seconds, stir the suspension again vigorously and allow to settle for 47 minutes 20 seconds (at 25°C); pipette a 25-ml sample at a depth of 10 cm into a weighed evaporating dish, and treat as in (ix) above. The difference in weight (wt soil + dish − wt dish) is $\frac{1}{40}$ of the weight of fine silt and clay in the original sample of soil. Find the weight of coarse silt.

(xi) Stir the suspension again vigorously and allow to settle for 7 hours 6 minutes (at 25°C); pipette a 25-ml sample at a depth of 10 cm into a weighed evaporating dish and treat as in (ix) above. The difference in weight (wt soil + dish − wt dish) is $\frac{1}{40}$ of the weight of clay in the original sample of soil. Find the weight of fine silt.

(xii) When the sedimentation tests have been completed wash the remaining solution clean of all fine particles leaving the sand fraction for sieving. This can be done by sieving and washing the sample through a 0·02-mm (limit of silt) mesh. This may prove difficult as the mesh holes may be blocked by soil particles.

Alternatively, therefore, the siphon method can be used. After completing operation (xi) above, siphon off the suspension to about 10 cm above the bottom of the cylinder. Transfer the residue into a 400-ml conical beaker calibrated at 10 cm. Make up to mark with distilled water. Stir vigorously and allow to settle for 4 minutes 15 seconds (at 25°C). Siphon off the supernatant liquid. Repeat this process until the supernatant liquid is clear.

The sand residue is then oven-dried at a temperature of 105°C after which it is put in a nest of sieves to obtain the different size-grades.

TABLE A.1
Table of Settling Times for Separation of Fractions

Temperature	To separate coarse silt ($20\,\mu$)		To separate fine silt ($6\,\mu$)		To separate clay ($2\,\mu$)	
°C	min.	sec.	min.	sec.	hr.	min.
10	6	14	69	13	10	23
11	6	3	67	20	10	6
12	5	54	65	27	9	49
13	5	44	64	40	9	34
14	5	35	62	7	9	19
15	5	27	60	30	9	5
16	5	19	59	0	8	51
17	5	10	57	24	8	37
18	5	3	56	0	8	24
19	4	55	54	36	8	12
20	4	48	53	20	8	0
21	4	41	52	0	7	48
22	4	28	50	54	7	37
23	4	24	49	36	7	26
24	4	22	48	24	7	16
25	4	15	47	20	7	6

Calculations:
Add together all the weights of the different size-grades from clay to coarse sand. Then express each weight as a percentage of the total weight of the fractions (i.e. weight of soil after removal of organic matter and soluble salts). The following is a step-by-step description of calculations done in a sedimentation analysis using an ordinary pipette of capacity 25 ml.

1 *Coarse sand + fine silt + clay*
   Wt dish + oven-dried 25-ml sample   = 25·26818 g
   Wt dish   = 24·84410 g
   Wt oven-dried 25-ml sample   = 0·42408 g
∴ Wt coarse sand + fine silt + clay in original sample = 0·42408 × 40 g
                                                                            = **16·96 g**

2 *Fine silt + clay*

Wt dish + oven-dried 25-ml sample	= 27·37490 g
Wt dish	= 27·04329 g
Wt oven-dried 25-ml sample	= 0·33161 g
∴ Wt fine silt + clay in original sample	= 0·33161 × 40 g
	= **13·26 g**

3 *Clay*

Wt dish + oven-dried 25-ml sample	= 26·09909 g
Wt dish	= 25·94050 g
Wt oven-dried 25-ml sample	= 0·15849 g
∴ Wt clay in original sample	= 0·15849 × 40 g
	= **6·34 g**

Hence,

Wt coarse silt = 16·96 − 13·26 g	= **3·70 g**
Wt fine silt = 13·26 − 6·34 g	= **5·92 g**

Table of Results:

Grade	Wt (g)	% of total
Coarse sand	2·58	11·40
Medium sand	0·58	2·56
Fine sand	3·41	15·07
Coarse silt	3·70	16·35
Fine silt	5·92	26·16
Clay	6·34	29·03
Total	22·63	99·57

Presentation of Results:

The table of results as shown above is one form of presentation. Other forms are:

(a) Summation curve (graph method) involving the plotting of the summation percentage P as a continuous function of x (or of $\log x$). The particles are assumed to be infinite in number and particle size to be a continuous variate.

(b) Block diagrams, in which the percentages within each size-grade are represented by proportional rectangles.

(b) *The hydrometer method*

Although it is less refined and less accurate than the pipette method, the hydrometer method is a simpler and less laborious method of particle-size analysis. This largely accounts for its popularity among engineers, geologists, and geographers. The principle of this technique is basically the same as that of the pipette sampling method. The difference lies in the way the concentration of soil particles in suspension is determined. Also, much larger quantities of soil than are used in the pipette method tend to be used in the hydrometer method, e.g. 50–100 g weight of soil may be used.

Apparatus:
The apparatus for pre-treatment and dispersion is the same as those given for the pipette sampling method. For the sedimentation tests the equipment needed is the 1,000-ml graduated measuring cylinder, the hydrometer, and a stop clock.

Procedure:
The process is the same as for the pipette sampling method up to the point where the soil suspension is transferred to a 1,000-ml cylinder after dispersion with 0·1 per cent calgon; see (vii) above. The cylinder is commonly stood in a water bath to maintain a constant temperature. The other procedures are:

(i) Clean the hydrometer and dry it. Take the temperature of the suspension; put a rubber stopper over the end of the cylinder and shake the contents thoroughly.
(ii) Start the stop clock immediately and slowly insert the hydrometer in the suspension. Ensure that the hydrometer is standing up as upright as possible in the suspension.
(iii) Take readings on the hydrometer at 30 seconds, 5 minutes, and after 2 hours. After each reading the suspension is shaken thoroughly as described above before starting the stop clock for the next reading. The times at which the readings are taken normally depend on the size-grades of particles whose concentrations in the suspension are required. The three readings at 30 seconds, 5 minutes, and 2 hours will give, respectively, the concentrations of sand + silt + clay; silt + clay; and clay.

Correction to hydrometer readings:
The hydrometer readings need to be corrected for temperature and the weight of calgon used in the particle fractionation and sodium carbonate.

(i) To correct for temperature, add 0·3 g for every 1°C above 19·5°C and subtract 0·3 g for every 1°C below.
(ii) For weight of calgon and sodium carbonate, subtract 0·3 g.

The different grades of sand (fine, medium, coarse) may be determined by the processes described under the pipette sampling method.

6 Soil pH

Soil pH is one of the first properties to be analysed as soon as the soil samples have been brought in from the field. This is because, with time, and whatever the precautions, the soil is gradually decomposed and the pH alters, with time, and whatever the precautions. The determination of soil pH in the laboratory follows the same procedure as in the field; the only difference is that there is scope for greater accuracy and precision in the laboratory.

The soil-to-water ratio in the suspension to be used for a pH test is important because the dilution of the soil with distilled water always has the effect of increasing pH regardless of the initial pH value of the soil or of the distilled water. A soil-to-water ratio of 1 : 5 or 1 : 2·5 is commonly used. The 1 : 5 dilution, for example, could increase the pH value by over 1 pH unit and the ratio 1 : 2·5 may be preferred. As a precaution, a soil-to-water ratio of 1 : 1 has been adopted by some workers.

Soil pH also varies with the neutral salt concentration, nitrates and sulphates. For instance, soil pH decreases during the hot dry season when soluble salts accumulate in the soil. These are subject to leaching during the relatively cool rainy season when pH increases again. It was specifically to offset the influence of seasonal variations in soluble salt concentration that Schofield and Taylor (1955) proposed a method for the determination of soil pH in 0·01M $CaCl_2$. The pH is measured in a suspension in which the ratio of soil to calcium chloride is 1 : 2. In this way, it is claimed, the determination will better reflect the intrinsic characteristic of the soil than would the soil pH value measured in water. The pH value obtained by this method is virtually independent not only of the initial salt concentration but also of the soil-to-water ratio. It is also claimed (Peech, 1965) that in as much as 0·01M $CaCl_2$ solution is approximately equivalent to the total electrolyte concentration of the soil solution of a non-saline soil at optimum fieldwater content, the soil pH measured in 0·01M $CaCl_2$ (ratio 1 : 2) represents more nearly the pH of the soil solution under actual field conditions.

The method of soil pH determination in 0·01M $CaCl_2$ solutions is described below.

Apparatus and Reagents:
(i) 50-ml beaker
(ii) 0·01M $CaCl_2$ solution
(iii) pH meter

Procedure:
(i) Weigh 10 g of soil into a 50-ml beaker.
(ii) Add 20 ml of 0·01M $CaCl_2$ solution. Allow to stand for 30 minutes, stirring the suspension occasionally.

(iii) Let the suspension stand for a further 30 minutes to allow the fine soil particles to settle. Then determine the pH of the supernatant liquid using a pH meter.

7 Carbonate

The determination of the soluble carbonate (and bicarbonate) in the soil is carried out simply by titrating a soilwater extract with a weak acid. For the determination of carbonate-C the soil is treated with acid to decompose the $CaCO_3$ and the amount of carbon dioxide released is measured directly. There are many methods for the determination of carbonate content and carbonate-C: some are accurate, direct, and quantitative; others are less accurate, indirect, and semi-quantitative, but quick and easy to operate (see Allison and Moodie, 1965, pp. 1379–87). The method described here is of the latter type.

The acid-neutralization method

This measures the amount of acid neutralized when the soil is treated with it to decompose the carbonates. This is used as a rough index of the soil carbonate.

Apparatus and Reagents:
(i) Standardized 0·5N HCl
(ii) 0·25N NaOH
(iii) Phenolphthalein, 1 per cent in 60 per cent ethanol
(iv) Beaker and watchglass

Procedure:
(i) Take a 5–25 g soil sample and add 50 ml of standardized 0·5N HCl in a 150-ml beaker. Cover the beaker with a watchglass.
(ii) Boil the soil–acid mixture gently for 5 minutes and then allow to cool.
(iii) Filter the mixture and wash the soil thoroughly with water to remove excess acid.
(iv) Titrate the HCl in the filtrate with 0·25N NaOH using the phenolphthalein as indicator.

Calculations:
(i) Carbonate equivalent %
$$= \frac{\text{added HCl (in milliequivalents)} - \text{used (in milliequivalents) NaOH} \times 0.05 \times 100}{\text{Wt dry soil (g)}}$$

(ii) Carbonate-C %
$$= \frac{\text{HCl added} - \text{HCl used} \times 0.006 \times 100}{\text{Wt dry soil (g)}}$$

8 Organic Carbon (+Organic Matter)

Two methods are used in determining organic-C, (1) based on quantitative combustion procedures wherein C is determined as CO_2, and (2) based on the reduction of the dichromate ion $(Cr_2O_7^{2-})$ by organic matter wherein the unreduced $Cr_2O_7^{2-}$ is measured by titration. The combustion methods are time consuming and require complicated sets of apparatus; hence the titration methods are more commonly used. However, the titration methods have two disadvantages: (1) the oxidation of C is incomplete, thus necessitating the use of correction factors to bring the C-values obtained close to those of the combustion methods; (2) the C-values are subject to error because of the presence in the soil of other oxidizable substances such as chloride, iron, and manganese. Only one of the titration methods, the Walkley–Black method, will be described. The combustion and other titration methods can be found in standard texts on methods of soil analysis.

The Walkley–Black method

The organic matter in the soil is oxidized by normal potassium dichromate. The reaction is facilitated by the heat generated when concentrated sulphuric acid is added. The excess dichromate not used up in the oxidation is determined by titration with standard ferrous ammonium sulphate. The amount of organic-C oxidized is calculated from the amount of dichromate reduced. The value is expressed as a percentage of the dry weight of soil used.

Apparatus and Reagents:
(i)   500-ml conical flask, pipettes, burette
(ii) Potassium dichromate $(N.K_2Cr_2O_7)$
(iii) Concentrated (not less than 96 per cent) sulphuric acid (H_2SO_4)
(iv) Concentrated phosphoric acid (H_3PO_4)
(v)   0·5 per cent di-phenylamine
(vi) Ferrous ammonium sulphate $(O.4NFe(AM)SO_4.7H_2O)$

Procedure:
(i)   Grind the soil to pass through a 100-μ-mesh sieve. Then weigh into a 500-ml conical flask a sample containing 10–25 mg of organic-C. Usually the weight of soil needed is 1 g for a mineral soil and 0·5 g or less for a highly organic soil.
(ii) Add, from a burette, 10 ml $N.K_2Cr_2O_7$ and swirl the flask gently to disperse the soil in the solution.
(iii) Add, immediately, 20 ml of concentrated H_2SO_4 from a measuring cylinder. Shake the mixture and allow to stand for twenty minutes.
(iv) Dilute the mixture with distilled water to about 250 ml. Add 10 ml of concentrated H_3PO_4 and 1 ml of 0·5 per cent di-phenylamine. The colour is bluish at this point.

(v) Titrate the chromic acid not used up in the oxidization with $0.4N Fe(AM)SO_4 \cdot 7H_2O$ from a burette. The solution first changes colour to purple or deep blue and after a few more drops of titrant the colour flashes to green. Take the reading on the burette to find out the quantity of titrant used.

(vi) At the same time prepare a blank, adding all the reagents but with no soil. Titrate the blank. Usually the blank takes 25 ml of $0.4N Fe(AM)SO_4 \cdot 7H_2O$ to reach the green end-point.

Calculations:

(i) Percentage organic-C $= \dfrac{V_1 - V_2 \times 0.003}{W} \times 100 \times f$

where V_1 = volume $N.K_2Cr_2O_7$ (i.e. 10 ml)
V_2 = volume $N.Fe(AM)SO_4 \cdot 7H_2O$ used in titration

Since 10 ml $N.K_2Cr_2O_7$ = 25 ml $0.4N Fe(AM)SO_4 \cdot 7H_2O$ (blank), it means that
1 ml $N.Fe(AM)SO_4 \cdot 7H_2O = 2.5 N.K_2Cr_2O_7$

$$\therefore V_2 = \dfrac{\text{titration reading}}{2.5}$$

W = weight of air-dry soil

0.003 is derived from the fact that 1 ml $N.K_2Cr_2O_7$ = 3 ml carbon

f = correction factor (usually 1.33)

(ii) percentage organic matter = organic-C percentage × 1.724

9 *Total Nitrogen*

Nitrogen exists in the soil in both organic and inorganic forms. Most is in organic form; small quantities occur as ammonium and as nitrate. The determination of soil N is hampered by two major obstacles: (1) current knowledge about the forms of N in the soil is far from adequate, and (2) the total N content of soils is very small indeed, ranging from less than 0.02 per cent in subsoils to just over 2.5 per cent in peats.

Two methods are generally employed for the determination of total N: (1) the Dumas method, whereby the N is oxidized in a dry combustion chamber, and (2) the Kjeldahl method in which the N is oxidized and converted into ammonium with an acid, and the quantity of ammonium produced is then determined by distillation with an alkali. Like all combustion methods, the Dumas method is time consuming and complicated.

APPENDIX

The Kjeldahl method, on the other hand, is easy to operate, efficient, and far less time consuming.

The Kjeldahl method of determining total N

The process is in two stages. (1) The soil sample is digested to convert N to ammonium. This is done by treating the soil with concentrated H_2SO_4, the reaction being facilitated by adding K_2SO_4 which raises the temperature of digestion. Catalysts such as selenium (Se), mercury (Hg), and copper (Cu) are used to promote oxidation of organic matter. (2) The ammonium in the digest is determined by titrating with an alkali.

Apparatus and Reagents:
(i) 500 ml macro-Kjeldahl flasks
(ii) Macro–Kjeldahl digestion stand and distillation apparatus
(iii) Concentrated H_2SO_4
(iv) Catalyst mixture of Se, $CuSO_4.5H_2O$ and K_2SO_4.
(v) Boric acid solution, methyl red + methylene blue indicator
(vi) 45 per cent NaOH
(vii) Standard 0·05N HCl

Procedure:
(a) Digestion
(i) Weigh 10 g of air-dry soil into a dry 500-ml macro-Kjeldahl flask.
(ii) Add 20 ml of water and after swirling the flask for about five minutes allow to stand for a further thirty minutes. Then add 10 g K_2SO_4, 1 g $CuSO_4.5H_2O$, 0·1 g Se, and 30 ml concentrated H_2SO_4. Mix the solution by means of a swirling motion.
(iii) Heat the flask cautiously in a fume chamber, swirling the flask occasionally until the digest clears and turns light green or grey in colour. Continue heating for another hour from this stage (some analysts suggest five hours).
(iv) Allow the flask to cool. Then add slowly, and whilst shaking the flask, approximately 100 ml of tap water. Transfer the solution to a clean flask for distillation.

As much as possible, retain any sandy residue in the digestion flask during the transfer to prevent the particles causing severe bumping during distillation.

Wash the sandy residue with approximately 50 ml aliquots until 250–300-ml solution is obtained.

(b) *Distillation*
(v) Put 50 ml of 4 per cent boric acid (H_3BO_3) solution in a 500-ml conical flask on which the 150-ml level has been marked. Add 3 drops of mixed indicator.
(vi) Place the conical flask under the condenser of the distillation apparatus so that the end of the condenser is dipped in the solution inside.
(vii) Put a small piece of litmus paper into the flask containing the diluted digest. Add 125 ml of 45 per cent NaOH by pouring it carefully down the side of the flask so that the alkali reaches the bottom of the flask without mixing appreciably with the digest.
(viii) Attach the flask to the condenser and swirl it to mix the contents. The litmus paper should now show that the solution is alkaline.
(ix) Distil, taking care to regulate the heat to minimize frothing or bumping, until about 150 ml of distillate has been collected.
(x) Titrate the distillate with $0.05N.H_2SO_4$ or HCl. The colour change is from green, through greyish-blue, to pink.
(xi) Prepare a blank and titrate as above.

Calculations:
milliequivalents of N in the sample = ml $0.05N$ H_2SO_4 used − blank value × normality of H_2SO_4
total percentage N = ml $0.05N$ H_2SO_4 × normality × f
 f (10-g sample) = 0.14 (correction factor)

10 *Exchange Acidity*

Exchange acidity, which is also referred to as 'exchangeable hydrogen' or 'total titratable acidity', includes not only hydrogen ions but also aluminium ions in the soil colloid. The exchange acidity together with the total exchangeable bases in the soil constitute its cation exchange capacity (CEC). The method described below for the determination of exchange acidity is the 'residual-carbonate method'. The exchangeable H^+ and Al^+ ions are replaced by C^{++} when the soil is treated with a buffer solution containing a mixture of a weak acid and its salt. The amount of Ca^{++} is then determined by titration.

Apparatus and Reagents:
(i) Conical flasks
(ii) Whatman filter paper no. 32
(iii) Paranitrophenol buffer solution + $N/25$ $Ca(OH)_2$
(iv) Bromocresol green indicator
(v) $N/20$ HCl

Procedure:
(i) Weigh a representative sample of soil, say, 10 g or less, and transfer it into a conical flask.

(ii) Add 30 ml of paranitrophenol buffer solution. Place 30 ml of the solution in another flask (to be used as a blank). Stopper the flasks well and shake for an hour on a reciprocal shaker.
(iii) Filter the mixture and the blank through Whatman filter paper no. 32.
(iv) Titrate 20 ml of each filtrate with N/20 HCl using 3 drops of bromocresol green indicator.

Note: The colour at first is deep yellow-green due to the combined colours of the paranitrophenol and the bromocresol green. Near the end-point the paranitrophenol becomes colourless and the indicator pure green. A few more drops of acid turn it yellow. The end-point is reached very suddenly and great care is needed.

Calculations:
exchange acidity in ml per 100 g of soil

$$= (B - T) \times N \times \frac{100}{W} \times \frac{30}{20}$$

B = blank titration of 20-ml buffer solution
T = titration reading of 20-ml filtrate
N = normality of HCl
W = weight of soil

11 Exchangeable Bases

The total amount of exchangeable cations in the soil can be determined in two ways: (a) by a method which yields results combining all the elements; and (b) by summation of values obtained by individual determinations of Ca^{++}, Mg^{++}, K^+, and Na^+.

In the determination of exchangeable cation concentration in the soil, there are some possible sources of error. When the soil is leached, cations derived from salts normally dissolved in soil water are also leached along with those from the clay-humus colloids. It is suggested that to reduce this kind of error the soil should first be washed with distilled water or aqueous alcohol to remove the salts. In acid soils, found in humid regions, such cations from soil water are very small in quantity. Secondly, free Ca and Mg, carbonate and gypsum, may be present in the soil and these can be dissolved, however slightly, by the leachate (usually ammonium acetate ($N.NH_4AC.$) or sodium hydroxide (NaOH)). Finally, an error in the determination may occur if the leaching of the soil is not done thoroughly and efficiently. With very acid soils with very low concentration of Ca^{++}, the leaching is easily and quickly done. The leachate is added to the soil and the solution is shaken very well on a reciprocal shaker and then centrifuged. This can be repeated once or twice to ensure maximum leaching. But with more calcareous soils the leaching is done more slowly and painstakingly.

The leaching can be done in two ways, both based on the same principles. The aim in both cases is to ensure that the leachate soaks through the soil very slowly and that it is thereby able to dissolve and remove all the metallic cations in the soil.

Method 1: by filtration
  (i) Weigh 25 g of air-dry soil into a 250-ml beaker. Add approximately 50 ml of neutral ammonium acetate solution. Stir very well and allow to stand for one hour.
  (ii) Transfer the soil to a Buchner funnel fitted with a moist Whatman filter paper no. 42 under gentle suction. Collect the leachate in a 250-ml filter flask. Continue to leach the soil slowly with small quantities of ammonium acetate until approximately 200 ml of leachate has been collected.

Method 2: the leaching column
The 'leaching column' consists of a long (about 1 m) glass tube open at both ends. The top is shaped like a funnel. At the bottom end is a rubber stopper with a hole in the middle through which a glass tubing is inserted. On top of the rubber stopper there is a wire or nylon mesh, fine enough to prevent soil particles from blocking the glass tubing through which the leachate drips into the conical flask underneath.

  (i) Set up the leaching column as described above.
  (ii) Put a small quantity of fine-grained acid-washed sand in the column to a depth of about 12 mm (to prevent fine soil particles blocking the mesh and the glass tubing).
  (iii) Weigh out 25 g of soil and mix it with about an equal weight of the acid-washed sand.

  If the soil is clayey or particularly fine-grained, the sand will make it more permeable. If the soil, on the other hand, is coarse and the leachate would soak through too quickly, the fine sand gives the soil more cohesiveness and makes it less permeable.

  Pour the mixture (soil + acid-washed sand) in the column. Tap the tube gently to ensure good packing.
  (iv) Put a small quantity of the acid-washed sand on top of the soil mixture.
  (v) Fill a 200-ml or 250 ml volumetric flask with ammonium acetate. Quickly invert the flask into the leaching column.

  The ammonium acetate soaks through the soil column and the leachate is collected in the volumetric flask underneath. The flask is kept in place until no leachate is dripping through the glass tubing. The leachate should be filtered if it is cloudy in which case it contains soil particles.

APPENDIX

DETERMINATION OF TOTAL EXCHANGEABLE METALLIC CATIONS

An aliquot of the leachate is evaporated so that the ammonium acetate is volatilized leaving the metallic cations in the acetates. These are ignited to convert acetate residue to oxides and carbonates. These are then dissolved in an excess of standard HCl, which is back-titrated with standard ammonium hydroxide solution. The difference between quantity of HCl added and the AmOH needed for back titration is a measure of the exchangeable metallic cation concentration in the soil.

Apparatus and Reagents:
(i) Evaporating dish and watchglass
(ii) Standard 0·1N HCl
(iii) 0·1N ammonium hydroxide solution
(iv) Dimethyl yellow indicator

Procedure:
(i) Transfer 150 ml of the leachate to an evaporating dish and evaporate to dryness, first on a low bunsen burner flame, and finally on a water bath.
(ii) Ignite the residue at a full red heat for 15–20 minutes until the ash is practically free from CO_2. Cool the dish slowly by gradually reducing the flame.
(iii) Pipette into the dish 50 ml standard 0·1N HCl. Cover the dish with a watchglass and warm over a very small flame to hasten the solution of the carbonates. Finally bring just to boiling to drive off the CO_2.
(iv) Standardize the 0·1N solution of AmOH against HCl using dimethyl yellow as indicator. Do the titration in a silica dish with, say, 50 ml HCl.
(v) Add dimethyl yellow to the solution from the soil and back-titrate with the standardized AmOH.

Calculations:
The total exchangeable metallic cations are calculated and expressed in milli-equivalents per 100 g of soil. In the experiment 150 ml of leachate was used corresponding to 15 g of soil (out of a total of 25 g).

∴ exchange metallic cations m.e./15 g $= (B-T)N$

B = ml AmOH needed to neutralize 50 ml HCl
T = ml standardized AmOH needed for back titration
N = normality of AmOH

∴ total exchange metallic cations, m.e./100 g soil $= (B-T)N \times \dfrac{100}{15}$

INDIVIDUAL DETERMINATION Ca^{++}, Mg^{++}, K^+, Na^+

There are a variety of approaches to the determination of Ca^{++}, Mg^{++}, K^+, and Na^+ in the soil. These include (a) atomic absorption spectrophotometry, (b) flame photometry, and (c) complexometric titration. The principles of the first two are described briefly while the third method is treated in greater detail.

(a) Absorption spectrophotometry

This method is relatively new and is based on the use of a sophisticated precision instrument, the absorption spectrophotometer. The spectrophotometer emits radiant energy of continuously variable wavelength. When a beam of this electromagnetic radiation passes through a soil extract the intensity of the beam is decreased. This reduction in intensity of light is related to the concentration of absorbing substances in the solution. The absorption spectrophotometer enables the selection of a suitable wavelength for the measurement of each particular element in the solution. The absorption spectrophotometer is particularly recommended for the divalent elements Ca^{++} and Mg^{++} to provide good sensitivity and reduce the amount of interference from other elements. The determination of Ca^{++} in particular is subject to interference by phosphates and aluminium. The effect of phosphates is reduced when lanthanum chloride is added to the soil leachate before determination.

(b) Flame photometry

The flame photometer has simplicity to its advantage and it has become, in recent years, one of the most widely used analytical methods. The technique is based on the fact that a large number of elements, when excited in a flame, emit radiation at characteristic wavelengths. The magnitude of the emission at the characteristic wavelengths is determined by the concentration of the element under investigation. This makes quantitative analysis of elements by flame photometry possible. Radiation intensities are measured directly by a photosensitive cell and the current produced is indicated directly on a meter. There is a photo-cell for each element capable of effectively isolating the radiation characteristic of that element. The flame must be regulated to suit the element being determined. The monovalent Na^+ and K^+ require a low-temperature flame since ionization is less and there is less interference from other elements. The divalent Ca^{++} and Mg^{++} require a high-temperature flame. This is so because of greater interference from other elements which decreases as temperature increases.

The flame photometer readings are converted to parts per million (ppm) values of the elements from pre-prepared standard curves. The standard curves are drawn from the readings on the meter of standard solutions containing, for example, 0, 20, 40, 60, 80, and 100 ppm of Ca^{++},

Mg^{++}, K^+, or Na^+ as the case may be. Readings from the graphs are converted to milli-equivalents by multiplying the values by the equivalent weight of each element, e.g. Ca^{++}: 40; Mg^{++}: 24·32; Na^+: 23; K^+: 39·1.

For the determination of these elements either by the absorption spectrophotometer or the flame photometer, the ammonium acetate leachate needs no further treatment. All that is required is to put a small quantity of the leachate in a cell and add a few drops of lanthanum chloride to prevent phosphate interference. The needle of the photometer is dipped in the solution and the reading is read directly from the meter.

(c) Complexometric titration of exchangeable Ca^{++} and Mg^{++}

This method is based on the fact that salts of EDTA (ethylene diamine-tetra acetic acid) or varsenate solution form complexes with ions of most of the metallic elements. It is a quick and reliable method of determining Ca^{++} and Mg^{++} concentration in the soil.

Apparatus and Reagents:
(i) 250-ml conical flasks
(ii) Standard N/50 varsenate, Ca^{++} and Mg^{++} solutions
(iii) 20 per cent KOH (potassium hydroxide)
(iv) Concentrated NH_3 (nitric acid)
(v) Calcein indicator
(vi) EBT indicator

Procedure:
exchangeable calcium + magnesium
(i) Place 50 ml of distilled water and 25 ml of concentrated NH_3 in a 250-ml conical flask.
(ii) Add 10 drops of Ca^{++} and Mg^{++} solution, and 2 drops of EBT indicator.
(iii) Titrate the solution with varsenate drop by drop to a definite end-point.
(iv) Then, immediately, add 20 ml of the ammonium acetate leachate (see above) and titrate to the same end-point as before.
Note: Colour change is from 'wine-red' to a clear blue.

exchangeable calcium only
(i) Place 50 ml of distilled water and 20 ml of 20 per cent KOH in a 250-ml conical flask.
(ii) Add 10 drops of a dilute solution containing Ca^{++} and Mg^{++}, and approximately 0·05 g calcein indicator.
(iii) Titrate with varsenate drop by drop to a definite end-point.
(iv) Add 20 ml of the ammonium acetate leachate (see above) and titrate with varsenate to exactly the same end-point as before.

Calculations:
1 ml N/50 varsenate = 0·02 mg-equivalent Ca^{++} or Mg^{++}
i.e. $0·02 \times 20$ mg Ca^{++} or $0·02 \times 12$ mg Mg^{++}

In cation exchange studies the results are given in terms of mg-equivalents of Ca^{++} or Mg^{++} per 100 g of air-dry soil.

(1) *Calcium*
wt of soil = 10·00 g
vol. of leachate = 250 ml
vol. of aliquot used in titration = 20 ml which corresponds to 0·800 g of soil.
Then if V_1 = volume of varsenate required, amount of Ca^{++} in soil = $0·02 \times V_1 \times 100/0·800$ mg-equivalents per 100 g dry soil.

(2) *Calcium + Magnesium*
If V_2 = volume of varsenate required, then amount of Mg^{++} in soil = $0·02 \times (V_2 - V_1) \times 100/0·800$ mg-equivalents per 100 g.

13 Cation Exchange Capacity

For the determination of CEC we have to go back to our leaching column (experiment 12). It will be remembered that the soil was leached with ammonium acetate to remove the exchangeable metallic cations. What has happened is that as each cation was removed from the negative exchange sites, its place was taken by NH_4^+. Thus to find the value for CEC the amount of NH_4^+ adsorbed by the soil must be determined. To do this it is necessary first to wash the soil clean of excess ammonium acetate. This is accomplished by leaching the soil with 95 per cent ethyl alcohol (1·0H). The soil is then leached again with acidified NaCl solution to dissolve the adsorbed NH_4^+, the quantity of which is determined by distillation on a Markham Still.

Apparatus and Reagents:
(i) Markham Still
(ii) 50 per cent sodium hydroxide (NaOH)
(iii) Boric acid indicator (0·25 per cent boric acid containing 45 ml methyl red + 15 ml bromocresol-green/2 litres)
(iv) N/50 HCl

Procedure:
(i) Transfer an aliquot of 10 ml of the NaCl leachate into a Markham Still. Add 6 ml of 50 per cent NaOH and 10 ml of boric acid indicator.
(ii) Distil for three minutes.
(iii) Titrate the distillate with N/50 HCl.
Note: The amount of HCl used is equivalent to the exchangeable ammonium adsorbed by $\frac{1}{25}$ (10 ml/250 ml) of the original weight of soil.

Calculations:

$$\text{CEC m.e./100 g soil} = \frac{(T-B) \times 0.02 \times (25) \times 100}{W}$$

T = titration reading
B = blank

$0.02 = 1$ ml N/50 HCl $= \frac{1}{50} = 0.02$

the factor (25) depends on original volume of leachate.

W = weight of soil

CEC — DETERMINATION BY SUMMATION
The above section describes how CEC can be determined directly in one operation. But it has been shown that CEC consists of two components: exchange acidity and total exchangeable bases. Thus one way of finding the value for CEC is to determine each of these components separately and then find their sum.

CEC = exchange acidity + total exchangeable bases

14 *Free Iron Oxide Fe_2O_3*
A common method for the determination of free iron oxide is the 'sodium-hydrosulphide method'.

Apparatus and Reagents:
(i)   100-ml centrifuge tube
(ii) 100-ml volumetric flasks
(iii) Conical beakers
(iv) 100-vol. (30 per cent) hydrogen peroxide (H_2O_2)
(v)   Solid sodium dithionite ($Na_2S_2O_4$)
(vi) N.Na_2SO_4 in N/1,000 H_2SO_4 (sulphuric acid)
(vii) Concentrated HNO_3 (nitric acid)
(viii) 2N ammonia
(ix) 10 per cent H_2SO_4
(x)   25 per cent thioglycollic acid
(xi) 10 per cent ammonium hydroxide (NH_4OH)

Procedure:
(i)   Take 1·0 g soil sample and grind to pass through a 0·5 mm sieve.
(ii) Place the sample in a beaker and add 20 ml of 100-vol. (30 per cent) H_2O_2. Evaporate to dryness.
(iii) Transfer into 100-ml centrifuge tube. Add 20 ml of distilled water and heat on a water bath to 40–45°C.

(iv) Add 0·8 g solid $Na_2S_2O_4$ fairly accurately. Prepare a blank with 50 ml of distilled water and 0·8 g $Na_2S_2O_4$. Keep at 40–45°C for forty minutes with occasional shaking.

(v) Centrifuge the solutions and wash twice with approximately $N.Na_2SO_4$ in $N/1000$ H_2SO_4. Collect the washings. Make up to 100 ml with distilled water in a 100 ml volumetric flask.

(vi) Put an aliquot of 25 ml in a conical beaker and evaporate with 8 ml of concentrated HNO_3 and 8 ml of 10 per cent H_2SO_4 until the liquid goes oily and fumes. Do the evaporation in a fume chamber.

(vii) Cool and add 50 ml of distilled water. Boil to dissolve the residue. Make up to 200 ml.

(viii) Pipette 25 ml aliquot into a 100-ml volumetric flask. Add 2N ammonia from a burette until the solution is just alkaline (use litmus paper as indicator).

(ix) Then, add *in order*: 5 ml of 20 per cent citric acid; 2 ml of 25 per cent thioglycollic acid; and 20 ml of 10 per cent w/v NH_4OH. Dilute to 100 ml with distilled water.

(x) Measure the colour intensity on the EEL Absorptiometer in a 1-ml cell, using a 604 filter. Use the blank to zero the meter.

Calculations:

The milligrams of Fe^{++} in 100-ml solution are read off a standard curve.

$$Fe_2O_3 = \text{milligrams of } Fe^{++} \text{ original solution} \times \frac{80}{56}$$

(a) The factor 80 is $\frac{1}{2}Fe_2O_3$ (112 + 48 = 160). The value is halved because of the divalency of Fe^{++}.

(b) The denominator 56 is the atomic weight of iron.

REFERENCES

Ajaegbu, H. I. and Faniran, A. (1973), *A New Approach to Practical Work in Geography*, Heinemann Educational Books, Ibadan.

Akroyd, T. N. W. (1957), *Laboratory Testing in Soil Engineering* (Soil Mechanics Ltd, Chelsea), Marshall Press Ltd, London.

Allen, V. T. (1948), 'Formation of bauxite from basaltic rocks of Oregon', *Econ. Geol.*, 43, 619–29.

Areola, O. O. (1971), 'Soil mapping from aerial photographs in Montgomeryshire, Wales', unpubl. PhD thesis, University of Cambridge, England.

— — (1974), 'Photo-interpretation of land facets as a soil mapping technique', *Geoforum*, 20, 25–38.

Avery, B. W. (1956), 'A classification of British Soils', *Trans. 6th Intern. Congr. Soil Sci.*, V, 279–85.

Bawden, M. G. and Tuley, P. (1966), *The Land Resources of Southern Sardauna and Southern Adamawa Provinces, Northern Nigeria*, Land Resources Study no. 2, LRD, Ministry of Overseas Dev., Surrey.

Beckett, P. H. T. (1968), 'Method and scale of land resources surveys in relation to precision and cost', in Stewart, G. A. (ed.), *Land Evaluation: Papers of a CSIRO Symposium*, 53–63, Macmillan, Australia.

— — and Webster, R. (1965), *A Classification System for Terrain*, MEXE Report no. 872, An Interim Report.

Black, C. A. (ed.) (1965), *Methods of Soil Analysis. Part I: Physical and Mineralogical Properties, etc. Part II: Chemical and Microbiological Properties.* American Society of Agronomy, Madison, Wisconsin, USA.

Blackwelder, E. (1933), 'The insolation hypothesis of rock weathering', *Amer. J. Sci.*, 26, 97–113.

Bowman, I. (1914), 'Forestry physiography, physiography of the United States and principal soils in relation to forestry', quoted in Heath, G. R. (1956), *Photogram. Eng.*, 22, 144–68.

Bridges, E. M. (1970), *World Soils*, Cambridge University Press.

Bruckner, W. D. (1955), 'The mantle rock ("laterite") of the Gold Coast and its origin', *Geol. Runds.*, 43, 307–27.

— — (1957), 'Laterite and bauxite of West Africa as an index of rhythmical climatic variations in the tropical belt', *Ecolog. Geol. Helv.*, 50, 239–56.

Buckman, H. C. and Brady, N. C. (1966), *The Nature and Properties of Soils*, 2nd edn, Macmillan, New York.

Bunting, B. T. (1965), *The Geography of Soil*, Hutchinson, London.

Buringh, P. (1954a), 'The analysis and interpretation of aerial photographs in soil survey and land classification', *Netherlands J. Agric. Sci.*, 2, 16–26.

— — (1954b), 'The analysis of pedological elements in aerial photographs', *Trans. 5th Intern. Congr. Soil Sci.*, Leopoldville, II, V–16.

— — (1955), 'Some problems concerning aerial photo-interpretation in soil survey', *Netherlands J. Agric. Sci.*, 3, 100–5.

— — Steur, G. G. L., and Vink, A. P. A. (1962), 'Some techniques and methods of soil survey in the Netherlands', *Netherlands J. Agric. Sci.*, 10, 157–172.

Burke, K. and Durotoye, B. (1970), 'The quaternary in Nigeria', *Bull. Assoc. Senegal et Quat. Ouest Afrique*, Dakar, 27, 70–95.
—— and Whiteman, A. J. (1971), 'A dry phase south of the Sahara 20,000 years ago', *Nig. J. Anthrop.*, 1, 1–8.
Butler, B. E. (1958), *Depositional Systems in the Riverine Plains in Relation to Soils*, CSIRO Soil Publication no. 10.
—— (1959), *Periodic Phenomena in Landscape as a Basis for Soil Studies*, CSIRO Soil Publication, no. 14.
Christian, C. S. (1957), 'The concept of land units and land systems', *Proc. 9th Pacific Sci. Cong.*, 20, 74–81.
Clarke, G. R. (1957), *The Study of the Soil in the Field*, 4th edn, Oxford University Press.
Cole, G. A. J. (1908), 'The red zone in the basaltic series of the county of Antrim', *Geol. Mag. new ser. dec.*, V, 5, 341–4.
—— et al. (1912), *The Interbasaltic Rocks (Iron Ores and Bauxite) of North-East Ireland*. Mem. Geol. Serv. Ireland, HMSO, Dublin.
Corbett, J. R. (1969), *The Living Soil: The processes of soil formation*, Martindale Press, NSW, Australia.
Crocker, R. L. (1952), 'Soil genesis and pedogenic factors', *Quart. Rev. Biol.*, 27, 139–68.
Cruickshank, J. G. (1972), *Soil Geography*, David & Charles, Newton Abbot.
Curtis, L. F., Doornkamp, J. C., and Gregory, K. J. (1965), 'The description of relief in field studies of soils', *J. Soil Sci.*, 16, 16–30.
Daubenmire, R. F. (1964), *Plants and Environment*, 4th edn, Wiley, New York.
Day, P. R. (1965), 'Particle fractionation and particle size analysis', in Black, C. A. (ed.), *Methods of Soil Analysis. Part I: Physical and Mineralogical properties, etc.*
D'Hoore, J. L. (1964), *Soil Map of Africa*, Scale 1 : 5,000,000, CCTA, Lagos.
—— (1968), in Moss, R. P. (ed.), *The Soil Resources of Tropical Africa*, Cambridge University Press, Part I, Chapter 1, pp. 7–28.
Dickson, B. A. and Crocker, R. L. (1954), 'A chronosequence of soils and vegetation near Mt Shastra, California. Pt III: some properties of the mineral soils', *J. Soil Sci.*, 5, 173–91.
Dury, G. H. (1969), 'Rational descriptive classification of duricrusts', *Earth Sci. J.*, 3(2), 17–36.
—— (1971), 'Relict deep-weathering and duricrusts in relation to the palaeo-environments in middle latitudes', *Geog. J.*, 137(4), 511–22.
—— and Knox, J. C. (1971), 'Duricrusts and deep-weathering profiles in southwestern Wisconsin', *Science*, 174, 291–2.
Ebisemiju, S. F. (1976), unpubl. Ph.D. thesis, University of Ibadan.
Ellison, W. D. (1947), 'Soil erosion studies—Part IV: Soil erosion, soil loss, and some effects of soil erosion', *Agric. Eng.* (July), 297–300. (See also articles by same author and in same journal for April, May, June, August, and by Ellison, W. D. and Ellison, O. T. in the same journal for September (Part VI, 402–405) and October (Part VII, 442–4).)
Faniran, A. (1968), 'A deeply weathered surface and its destruction', unpubl. Ph.D. thesis, University of Sydney, Australia.
—— (1969), 'Duricrust, relief and slope population in the Sydney district of

REFERENCES

New South Wales', *Nig. Geog. J.*, 12, 53–62.

—— (1970a), 'Maghemite in the Sydney duricrusts', *Amer. Mineral.*, 55, 925–33.

—— (1970b), 'Landform examples from Nigeria, No. 2; The deep weathering (duricrust) profile', *Nig. Geog. J.*, 13, 87–8.

—— (1971), 'Implications of deep weathering on the location of natural resources', *Nig. Geog. J.*, 14, 59–69.

—— (1974), 'The extent, profile and significance of deep weathering in Nigeria'. *J. Trop. Geog.*, 38, 19–30.

—— and Areola, O. O. (1974), 'Landform examples from Nigeria No. 7—A gully', *Nig. Geog. J.*, 17, 57–60.

Fenneman, M. N. (1916), 'Physiographic divisions of the United States', *AAAG*, 6, 19–98.

Floyd, P. (1965), 'Soil erosion and deterioration in Eastern Nigeria', *Nig. Geog. J.*, 8, 33–44.

Food and Agricultural Organization (UNO) (1960), *Soil Erosion by Wind and Measures for Its Control on Agricultural Lands*, Agric. Dev. Paper no. 71, Rome.

—— (1965a), *Soil Erosion by Water: Some Measures for Its Control on Cultivated Lands*, Agric. Dev. Paper no. 81, Rome.

—— (1965b), Agricultural Development in Nigeria 1965–1980, Rome.

—— (1965), *Guidelines for Soil Description*. Soil Survey & Fert. Branch, Land & Water Dev. Div.

Fox, C. S. (1932), *Bauxite and Aluminous Laterite*, 2nd edn, Crosby Lockwood, London.

Gerasimov, I. P. and Glazovskaya, M. A. (1965), *Fundamentals of Soil Science and Soil Geography*, Israel Program for Scientific Translation, Jerusalem.

Goldich, S. S. (1938), 'A study of rock weathering', *J. Geol.*, 46, 17–58.

Gordon, M., Jr, Tracey, J. I., and Ellis, W. M., (1958), *Geology of the Arkansas Bauxite Region*, US Geol. Surv. Prof. Paper 299.

Griggs, D. T. (1936), 'The factor of fatigue in rock exfoliation', *J. Geol.*, 44, 781–96.

Grove, A. T. (1951), *Land Use and Soil Conservation in parts of Onitsha and Owerri Provinces*, Geol. Surv. Nig. Bull. no. 21.

—— (1952), *Land Use and Soil Conservation on the Jos Plateau*, Geol. Surv. Nig. Bull. no. 22.

Grubb, P. L. C. (1965), 'Fuse laterite in a caldera near Skimpton, western Victoria, *Pro. Ro. Soc. Vic.*, 78(ii), 197–9.

Gunn, R. H. (1967), 'A soil catena on denuded laterite profiles in Queensland', *CSIRO Aust. Div., Land Res. Reg. Surv. Ann. Rept* for 1965–6, 17.

Harder, E. C. (1952), 'Examples of bauxite deposits illustrating variations in origin', in *Problems of Clay and Laterite Genesis*, Amer. Inst. Min. Metall. Engrs, 35–64.

Harrison, J. B. (1933), *Katamorphism of Igneous Rocks under Humid Tropical Conditions*. Imperial Bur. Soil Sci., Harpenden, Herts.

Hays, J. (1967), 'Land surfaces and laterites in the north of the Northern Territory', in Jennings, J. N. (ed.), *Landform Studies from Australia and New Guinea*, Mabbutt & ANV Press, Canberra, 182–210.

Jenny, H. (1941), *Factors of Soil Formation*, McGraw-Hill, New York.

Joffe, J. S. (1949), *Pedology*, Rutgers University Press, New Brunswick, NJ.
Jones, R. G. B. (1958), 'The application of air photo-analysis and interpretation to soil survey and land classification', *Rhod. Agric. J.*, 55(2), 195–201.
Keller, W. D. (1957), *Principles of Chemical Weathering*, Lucas Bros, Columbia, Missouri.
King, L. C. (1950), 'The study of the world plainlands: a new approach in geomorphology', *Quart. J. Geol. Soc.*, London, 106, 101–31.
Klinkenberg, K. (1967), *The Soils of the Lau-Kaltungo Area*, Bull. no. 36, Inst. for Agric. Res., Samaru, Zaria.
Kubiena, W. L. (1953), *The Soils of Europe*, Murby, London.
Lukashev, K. T. (1970), *Lithology and Geochemistry of the Weathering Crust*, Israel Program for Scientific Translation, Jerusalem.
Marbut, C. F. (1927), 'A scheme for soil classification', *1st Intern. Congr. Soil Sci. Proc.*, 4, 1–31.
— — (1935), Soils of the United States, in *Atlas of American Agriculture*, US Dept of Agriculture, Part III.
Merrill, G. P. (1897), *A Treatise on Rocks, Rock Weathering and Soils*, Macmillan, London.
Milne, G. (1935), 'Composite units for the mapping of complex soil associations, *Trans. 3rd Intern. Congr. Soil Sci.*, 1, 345–7.
Mohr, E. C. J. and Van Baren, F. A. (1959), *Tropical Soils*, NV Uitgeverij W. Van Hoeve, The Hague.
Moss, R. P. (1968), 'Soils, slopes and surfaces in tropical Africa', in Moss, R. P. (ed.), *The Soil Resources of Tropical Africa*, Cambridge University Press, 29–60.
— — (1969), 'The appraisal of land resources in tropical Africa: a critique of some concepts', *Pacific Viewpoint*, 10(2), 18–27.
Nikiforoff, C. C. (1931), 'History of A, B, C', *Amer. Soil Surv. Assoc. Bull.*, 12, 67–70.
— — (1942), 'Fundamental formula of soil formation', *Amer. J. Sci.*, 240, 846–866.
— — (1943), 'Introduction to palaeopedology', *Amer. J. Sci.*, 241, 194–200.
Nir, D. and Klein, M. (1974), 'Gully erosion induced by changes in land use in a semi-arid terrain', *Nahal Shigma, Israel Symposium Geomorphic Process in Arid Environments*, Israel.
Northcote, K. H. (1965), *A Factual Key for the Recognition of Australian Soils*, CSIRO Div. Soils, Report no. 2/65.
Norton, E. A. and Smith, R. S. (1930), 'The influence of topography on the soil profile character', *J. Amer. Soc. Agron.*, 22, 251.
Nye, P. H. and Greenland, D. J. (1960), *The Soil Under Shifting Cultivation*, Tech. Bull. no. 51, Commonwealth Bureau of Soils.
Ofomata, G. E. K. (1965), 'Factors of soil erosion in the Enugu area of Nigeria', *Nig. Geog. J.*, 8, 45–59.
Ollier, C. D. (1963), 'Insolation weathering: examples from central Australia', *Amer. J. Sci.*, 61, 376–87.
— — (1969), *Weathering*, Oliver & Boyd, Edinburgh.
Petersen, R. G. and Calvin, L. D. (1965), 'Sampling', in Black, C. A. (ed.), *Methods of Soil Analysis, Part I: Physical and Mineralogical Properties, etc.*, *J. Amer. Soc. Agron.*, 54–72.

Polynov, B. B. (1937), *Cycle of Weathering* (trans. A. Muir), Murby, London.

Reiche, P. (1950), *A Survey of Weathering Processes and Products*, University of New Mexico Press, Albuquerque.

Robinson, G. W. (1949), *Soils, Their Origin, Constitution and Classification*, Murby, London.

Rudeforth, C. C. (1969), 'Quantitative soil surveying', *Welsh Soils Disc. Group Rep. No. 10*, 42–8.

Russell, E. W. (1963), *Soil Conditions and Plant Growth*, Longmans, London.

Sherman, D. G. (1952), 'The genesis and morphology of alumina-rich laterite clays', in *Problems of Clay and Laterite Genesis*, Amer. Inst. Min. Metall. Engrs, New York.

Simonson, R. W. (1968), 'Concept of soil', *Adv. Agronomy*, 20, 1–47.

Smith, D. D. and Whitt, D. M. (1948), 'Estimating soil loss from field areas', *Agric. Eng.*, 29, 394–6.

— — and Weschmeier, W. H. (1957), 'Factors affecting sheet and rill erosion', *Trans. Geophys. Union*, 38, 889–96.

Smyth, A. J. and Montgomery, R. F. (1962), *Soil and Land Use in Central Western Nigeria*. Govt Printer, Ibadan.

Stace, H. C. T., Hubble, G. D., et al. (1968), *A Handbook of Australian Soils*, Rellim, South Australia.

Stallings, J. H. (1957), *Soil Conservation*. Prentice-Hall, Englewood Cliffs, NJ.

Stephens, C. G. (1946), *Pedogenesis Following the Dissection of Lateritic Regions in South Australia*, CSIRO, Austr. Bull. no. 206.

— — (1953), *Soil Surveys for Land Development*, FAO Agric. Studies no. 20, Rome.

— — (1962), *Manual of Australian Soils*, 3rd edn, Melbourne University Press.

Stewart, G. A. (1968), 'Land evaluation', in Stewart, G. A. (ed.), *Land Evaluation Papers of a CSIRO Symposium*, Macmillan, Australia, 1–10.

Sykes, R. A. (1940), 'A history of anti-erosion work at Udi', *Farm and Forest*, 1(1), 3–6.

Taylor, J. A. (1960), 'Method of soil study', *Geography*, XIV, 52–67.

Thomas, M. F. (1969), 'Geomorphology and land classification in tropical Africa', in Thomas, M. F. and Whittington, G. W. (eds), *Environment and Land Use in Africa*, Methuen, London, 103–45.

Thorp, J. and Smith, G. D. (1949), 'Higher categories of soil classification: order, suborder and great soil groups', *Soil Sci.*, 67, 117–26.

United States Department of Agriculture (1951), *Soil Survey Manual*, Agric. Handbook no. 18, Washington DC.

— — (1966), *Aerial Photo-Interpretation in Classifying and Mapping Soils*, Agric. Handbook no. 294, Washington DC.

Valette, J. and Higgins, G. M. (1967), *The Reconnaissance Soil Survey of an Area near Auna Niger Province, Northern Nigeria*, Soil Surv. Bull. no. 34, IAR, Samaru, Zaria.

Vilenskii, D. G. (1950), *Pochvovedene* (Soil Science), Uchpedgiz.

Watts, D. (1971), *Principles of Biogeography*, McGraw-Hill, London.

Wright, R. L. (1963), 'Deep weathering and erosion, surfaces in the Daly river basin, Northern Territory', *J. Geol. Soc.*, 10, 151–67.

Zingg, A. W. (1940), 'Degree and length of land slope as it affects soil loss in runoff', *Agric. Eng.*, 21, 59–64.

List of Abbreviations

AAAG	Annals of the Association of American Geographers
Agric. Eng.	Agricultural Engineering
Amer. Inst. Min. Metall. Engrs	American Institute of Mineral and Metallurgical Engineers
Amer. Mineral.	American Mineralogy
Amer. Soil Surv. Assoc. Bull.	American Soil Survey Association Bulletin
Amer. J. Sci.	American Journal of Science
CSIRO	Commonwealth Scientific and Industrial Research Organization
Earth Sci. J.	Earth Science Journal
Ecolog. Geol. Helv.	Ecologische Geologische Helvetia
Econ. Geol.	Economic Geology
Geog. J.	Geographical Journal
Geol. Mag. new ser. dec.	Geological Magazine new series
Geol. Runds.	Geologische Rundschau (Stuttgart)
Geol. Surv. Nig. Bull.	Geological Survey of Nigeria Bulletin
Geol. Surv. Prof. Paper	Geological Survey Professional Paper
Imperial Bur. Soil Sci.	Imperial Bureau of Soil Science
J. Agric. Sci.	Journal of Agricultural Science
J. Amer. Soc. Agron.	Journal of American Society of Agronomy
J. Geol.	Journal of Geology
J. Sci.	Journal of Science
J. Soil Sci.	Journal of Soil Science
J. Trop. Geog.	Journal of Tropical Geography
LRD	Land Resources Division Ministry of Overseas Development
Mem. Geol. Serv. Ireland	Memoire of the Geological Service of Ireland
MEXE	Military Engineering Experimental Establishment
Netherlands J. Agric. Sci.	Netherlands Journal of Agricultural Science
Nig. Geog. J.	Nigerian Geographical Journal
Nig. J. Anthrop.	Nigerian Journal of Anthropology
Pacific Sci. Cong.	Pacific Science Congress
Photogram. Eng.	Photogrammetric Engineering
Proc. Ro. Soc. Vic.	Proceedings of the Royal Society, Victoria
Quart. J. Geol. Soc.	Quarterly Journal of the Geological Society
Quart. Rev. Biol.	Quarterly Review of Biology
Rhod. Agric. J.	Rhodesian Journal of Agriculture
Trans. Geophys. Union	Transactions of the Geophysical Union
Trans.-Intern. Congr. Soil Sci.	Transactions of the 6th International Congress of Soil Science

INDEX

Ajaegbu, H. I., 97
Alkali cations, 182
Allen, V. T., 188
Aluminium, 10, 15, 21, 29, 43, 46, 72, 121, 151, 154, 163, 199
  hydrated and dehydrated forms, 11
  in conditions of extreme acidity, 11
Aluminium oxide, 39
Alumino-silicates, 180
Ammonia, 13
Aragonite, 43
Areola, O., 80, 90, 213
Argon, 14
Atterberg, A., 240
Avery, B. W., 137

Basalt, 44
Basalt soils, 4
Basement complex, 104
Bawden, M. G., 93
Beckett, P. H. T., 83
Black, C. A., 255
Blackwelder, E., 38
Blue-green algae, 200
Bowman, I., 89
Brady, N. C., 96, 192, 193
Bridges, E. M., 127, 155
Brown earths
  acid, 177
  characteristics of typical profile of, 177
  ferritic, 177
Bruckner, W. D., 191
Buchanan, Francis, Lord Hamilton, 161
Buckman, H. C., 96, 192, 193
Bunting, B., 8
Buringh, P., 81, 82
Burke, K., 164
Butler, B. E., 189, 191

Calcification
  calcic horizons, 179, 182
  formation of calcic and mollic horizons, 75
Calcite, 43, 203
Calcium, 10, 25
  chief exchangeable base, 11
  concentration as measure of degree of leaching, 203
  derived from, 11, 203
  role in, 203
Calcium bicarbonate, 75
Calcium carbonate, 11, 51, 121
Calcium sulphate, 11, 75
Calcrete, 46
  formation, 75
Carbon dioxide, 13, 14, 40, 45
Carbonic acid, 40
Carter, C. F., 93
Catena, 101, 185
Chalk, 37
Chernozem, best example of a mollisol, 75

Chernozem problem, 4
Christian, C. S., 90, 122
Clarke, G. R., 23, 105, 114, 122
Clay, 12, 108
Clay–humus complex, 27, 200, 203
Clay minerals
  and soil chemistry, 24, 27
  cation-exchange capacity, 25
  cation exchange and leaching, 26
  exchangeable hydrogen, 26
  formation and structure, 24
  illite, 26
  kaolin, 26
  montmorillonite, 26
  product of weathering, 46
  stages of weathering, 26
  total exchangeable bases, 26
Climax concept, 5
Cole, G. A. J., 189
Colloidal alumina, 46
Colloidal silica, 46
Colloids, 151
Community ecology, 89
Copper, 10
Corbett, J., 72, 132, 135, 150, 155
Cruickshank, J. G., 79, 139, 198, 203
Cryoturbation, 185
Curtis, C. F., 102

Darwin's evolutionary concept, 5
Davisian cycle concept, 5
Deep weathering profile, 161, 187–8
  formation of indurated, mottled, pallid zones, 72
  indurated zone, 40
  truncated or stripped, 56
D'Hoore, J. L., 95
Diorite soils, 4
Dokuchaev, V. V., 4–5, 7, 47, 48, 49, 58, 79, 99, 128
Doornkamp, J. C., 102
Durotoye, Bisi, 164, 191
Dury, G. H., 161, 188, 189

Edaphology, 237
Ellison, W. D., 207
Eluvial horizons, 67
Exchangeable hydrogen, 27
Experimental pedology, development of, 6

Faniran, A., 34, 56, 72, 97, 126, 161, 165, 213
Feldspars, 11, 40, 41
Fenneman, M. N., 89
Floyd, B., 223, 232
Food and Agricultural Organization, 6, 80
Fox, C. S., 189
Free alkali salts, 182

Free survey, 85
Frost shattering, 186
Fulvic acid, 173
  role in podzolization, 27

Gerasimov, I. P., 8, 78, 80
Gibbsite, 39, 40, 46, 165
Glazovskaya, M. A., 8, 78, 80
Gleization
  gleying, 76
  peat formation, 77
  surface- and ground-water gleying, 77
Goethite, 39, 40, 166
Goldich, S. S., 44
Gordon, M. J., 188
Granite soils, 4
Gravitational water, 15
Greenland, D. J., 204
Gregory, K. J., 102
Grid mapping, 83, 84
Griggs, D. T., 38
Grove, A. T., 216
Grubb, P. L. C., 161
Gunn, R. H., 56
Gypsum, 121

Haematite, 39, 40, 165
Harder, E. C., 188
Harrison, J. B., 169
Hays, J., 165
Hilgard, 79
Horizon differentiation, 106
Hornblende, 11
Hubble, 134
Humid temperate soils
  acid brown soil, 174, 175
  brown earths, 175
  grey-brown podzolics, 177–8
  podzolic soils, 73, 75, 76, 135, 173, 175
  podzols, 135, 154, 157, 173, 175, 176
  red and yellow podzolics, 135, 178–9
Humid tropical soils
  bauxite, 187, 189
  ferricretes, 155, 187
  ferrisols, 169
  krasnozems, 135, 170
  laterites, 135, 154, 155, 157, 158, 160, 176, 178, 187
  lateritic soils, 135, 157, 158, 160, 169, 176, 178
  latosols, 157, 160, 169
  oxisols, 157, 160
  red earth, 135, 171, 175
  silcretes, 155, 169, 187
  yellow earth, 135, 172, 175
Humification, 67
  effect of vegetation type, 69
Hydrogen sulphide gas, 42
Hydromorphic soils, 76, 77
Hydrous compounds, 39
Hydrous oxides, 21

Hydroxides of iron and aluminium, 161

Illuvial horizons, 68
Illuviation, 151
Inceptisol, 76
Institute for Agricultural Research, Samaru, Zaria, 213
International Institute of Tropical Agriculture (IITA), Ibadan, 158, 213, 214
Iron, 10, 15, 17, 18, 24, 29, 41, 43, 46, 72, 73, 117, 121, 151, 154, 163, 199
  hydrated and dehydrated forms, 11
  in conditions of extreme acidity, 11
Iron oxide, 39, 170

Jenny, H., 48, 49
Joffe, 7, 8
Jones, R. G. B., 82

K-cycle theory, 189
Kaolin, 39, 41
Keller, W. D., 40
King, L. C., 189
Klein, M., 231
Kosciusko uplifts, 187
Krasnozem, 72
  distinguishing characteristic, 171
  origin of term, 170
Kubiena, W. L., 137

Land capability classification, 145
Land Resources Division (British), 158
Landscape science
  concept of habitat-site, 90
  concept of integrated landscape, 90
  genetic landscape classification, 89
  land evaluation, 91
  land facets, 90
  land system, 90, 101
Laterite
  and lateritic soils, 71
  controversy, 160-1
  definition, 161
  depth measurements, 167, 168
  ferruginous tropical soil, 161
  formation of profile, 167, 169
  kongi/Agodi gravel, 164
  mineralogy, 161-3
  profile, 164-7
  secondary and reworked, 164
Lateritic soils
  ferralitic profile, 169-70
  reweathered laterites, 170
Lateritization
  alteration of high-silicate clays, 72
  mineralization of organic matter, 72
  process of desilication, 72
  regions where it operates, 71

stratified, unstratified profiles, 72
unique processes, 169
Leaching, 173
  of materials from A to B horizon, 29, 52
  on upper slopes, 59
  principal operating process in podzolization, 73
Lead, 10
Lepidochrocite, 166
Lime, 17
Lime soils, 3
Limestone, 37, 40, 203
Living organisms, importance in the soil, 13, 14
Living soil, 1, 65
Lukashev, K., 34

Macro-pores, 9
Maghemite, 165
Magnesium, 10, 25, 199
  deficiency in tropical soils, 204
  derived from, 11
  exchangeable base of, 11
  reserve in rock minerals, 204
Magnesium bicarbonate, 75
Manganese, 10, 17, 18, 29, 41, 46, 73, 121
Manganese duricrust, 161
Manganese oxide, 40, 161
Marbut, C. F., 6, 7, 128
Marcasite, 40
Marl soils, 3
Merrill, G. P., 39
Metallic cations, 10
Micas, 11, 17, 204
Micro-pores, 9, 10
Milne, G., 61
Mineral cycling, theory of, 3
Mohr, E. C. J., 160
Molybdenum, 10
Montgomery, R. F., 60, 91
Mor humus, 13
Moss, P. P., 91
Mottling, 59
Mull humus, 13

Natural resource inventory, 1
New American classification system
  diagnostic horizons, 138-9
  soil orders, 135-6
Nikiforoff, C. C., 63, 125
Nir, D., 231
Nitrobacter, 200
Nitrogen, 10
  cycle, 197
  fixation, 70, 200
  functions in plants, 201
  nitrate anion forming amino-acids, 201
  nitrification, 200
  susceptibility to leaching, 200
Nitrosomonas, 200
Northcote, K. H., 132
Norton, E. A., 58
Nye, P. H., 204

Ofomata, G. E. K., 216, 223
Olivine, 11
Ollier, C. D., 33, 44
Opal, 41
Orthoclase, 11
Oxygen
  oxidizing agent, 39
  presence in air and water, 39

Parent material, soils reflecting, 56-7
Particle-size analysis
  hydrometer method, 251-3
  pipette method, 247-51
  pre-treatment of samples, 246-7
Peat, 13
Pedalfers, 128, 179, 181, 182
Pedocals, 128, 179, 181, 182
Pedoclimate, 160
Pedogenesis, 1
Pedology, 237
  course, 8
  definition, 1
  history of development, 3-7
  relations of, 2
  Russian school of, 4
Pedology and geography, 1-2
  methodology of, 2
  unifying concepts, 2
Percolating water, 59
Permafrost, 76, 184, 185
Phosphorus, 10
  cycle, 200, 201
  inorganic pool, 201
  organic fertilizers, rich in, 201
  role in plant, 201
Physiographic mapping, 85
Plant materials, composition, 13
Plant roots
  factor of composition of soil air, 193
  factors which influence development of, 192
  optimum temperature for growth of, 193
  sensitivity to carbon dioxide, 193
  water supply to, 194
Playa, 183, 184
Podzol
  bleicherde, 173, 178
  groundwater, 174
  ideal soil profile, 73
  iron and humus iron, 174-5
  where best developed, 173-4
  word derived from, 173
Podzolization
  outcome of, 73, 75
  prevalence in cool humic regions, 73, 75
Pollution, and soil erosion, 204-5
Polynov, B. B., 33
Pore spaces, shapes of, 14
Porosity
  aggregate porosity defined, 116
  spaces between aggregates, 116-17

INDEX

Potassium, 10, 25
 added through mineral fertilizers, 204
 derived from, 204
 in ash from burnt vegetation, 204
 required by plants, 203
 sources of, 11
Potassium carbonate, 204
Potassium chloride, 204
Potassium sulphate, 204
Pseudomonas, 200
Pyrite, oxidation of, 40
Pyroxene, 11
Pyrrhotite, 40

Quartz, 10, 11, 17, 41, 44, 46
 oxide of silicon, 11
Quartzite, 44

Regosols, 191
Reiche, P., 33, 44
Relict soils
 buried soils, 189, 191
 podzolized laterites, 186
 relict duricrusts, 187–9
Rhizobium, 200
Robinson, G. W., 79, 128, 132
Rocks, endothermal sources of energy, 55
Rock weathering
 carbonation, 40
 chemical weathering, features of, 38
 colloidal plucking, 37, 38
 deep weathering, 34
 definitions, 33–4
 dissolution or solution, 41
 energy and, 45
 expansion factors of rock minerals, 36
 factors influencing, 42–5
 factors of, 34
 frost shattering, 37
 hydration, 39
 hydrolysis, 40–1
 interstitial air and water, role of, 35–6
 mechanical weathering, 37–8
 oxidation, 39
 product of, 45–6
 reduction, 42
 salt weathering, process of, 37
 supergenesis, 34
 thermal weathering, 35
 thermal weathering, extent of, 38
 types of weathering, 34–5
 weathering crust, 34
 weathering sequence of rocks, 43
Rothamsted Experimental Station, 79
Rudeforth, C. C., 84
Russell, E. W., 193, 198
Russian chernozem, 4
Russian Chernozem Commission for the Free Economic Society, 4
Russian period of pedology, 5–6
Rutile, 46

Salinization
 capillary action, 76
 related processes, 76
 solonchaks formed, 76
Sand, 11, 108
Sandstone, 37
Sandy soils, 11–12
Saprolite, 7, 45, 46, 65
Savanna, 163, 165, 181, 182
Schick, A. P., 36
Semi-arid and arid soils, *see* soils, semi-arid and arid
Serpentine, 39
Sesquioxide enrichment, 171
Sesquioxides, 106, 151, 164
Shaler, 79
Shifting cultivation, 169, 215
Sibirtzev, Nikolai, 4, 5, 128
Silica, 73, 161
 amorphous, 46
Silicate clays, 179, 180
 removal of, 151
Silicate minerals, 11
Silicon, 10, 43
Silt, 12, 108
Slope aspect, 59
Slope, factor of soil depth, 58
Smith, D. D., 58, 129, 211, 216
Smyth, A. J., 60, 91
Sodium, 204
Sodium chloride, 67, 154
Sodium sulphate, 37
Soil
 approaches to the scientific study of, 3
 as chemical laboratory, 3
 as collection of matter, 3
 as geological material, 4
 as medium for plant growth, 192
 components of the body, 9
 concepts of, 3
 definition and study of, 7–8, 9
 edaphological approach to study of, 192
 major element of land, 1
 mineral matter, 9, 10–12
 moisture-retentive capacity of, 199
 raw mineral, 191
 volume composition of, 9–10
Soil aeration, 10
 relation with oxidation–reduction processes, 15
Soil air
 composition, 14, 15
 diffusion in soil, 14
 diffusion of gases in soil, 195
 effects of poor aeration, 195
 gaseous exchange, 193, 194, 196
 good, defined, 193
 rate of carbon dioxide production, 194
 soil air–moisture ratio, 194–5
Soil analysis, methods of
 absorption spectrophotometry, 262
 Atterberg limits, 240
 bulk density, 242–3
 carbonate content, 254
 cation-exchange capacity, 264–5
 complexometric titration of metallic cations, 263–4
 exchange acidity, 258–9
 exchangeable bases, 239–4
 flame photometry, 262–3
 free iron oxide, 265–6
 laboratory equipment and organization, 238
 liquid limit test, 240–2
 moisture content, 238–9
 organic carbon, 255–6
 particle-size analysis, 246–52
 plastic limit test, 242
 size distribution of aggregates, 244–6
 soil pH, 253–4
 total nitrogen, 256–58
Soil capability, 3
Soil catena, 60, 61, 131
Soil chemistry, field of, 1
Soil classification, 3, 6, 237
 Australian systems, 132–5
 empirical or taxonomic approach, 125, 126
 for irrigation agriculture, 148–9
 for specific purposes, 145
 genetic approach, 125, 128–31
 great soil group, 130
 integrative approach, 125, 131
 irrigation suitability maps, 149
 morphological approach, 125, 127–8
 principles of natural distribution, 124–5
 soil association, 101, 130–1
 soil series, 6, 101, 111, 126, 127, 131
 systems in Europe, 136–7
 systems in United States, 142–3
 UNESCO/FAO system, 143
Soil colloids, 180
Soil colour
 description of, 17
 determining in the field, 107–8
 Munsell colour chart, 18
 Munsell colour system, 107
 problems of determination, 17–18
 producing substances, 17
Soil conservation
 approaches, 228–9
 contour farming, 229, 230, 231
 control of wind erosion, 234–6
 control of wind erosion on cultivated lands, 234–5
 control of wind erosion on grazing lands, 235–6
 in Enugu area (Nigeria), 232–4
 objectives, 228
 reclamation of badly eroded lands, 232

strip cropping, 229, 230, 231, 235
terrace agriculture, 229, 230, 231
water erosion control measures, 229–32
Soil constituents, 9
Soil drainage, 10
  aerobic, anaerobic conditions, 23
  classes, 23–4
  gleying, 117
  gleying, mottling, 24
  moisture-to-air ratio, 23, 117
  pattern of mottling, variables, 118
Soil erosion
  agents of, 218
  agriculture and, 213, 215–16
  and pollution, 204–5
  attributes of physiography affecting, 211
  by mining, 216
  by surface flow, 220–1
  by water, 218–25
  by water, factors of, 210
  by wind, 225–8
  defined as, 206
  detachability, transportability of soil particles, 207
  detaching, transporting capacity of agent, 207
  dynamic threshold velocity, 226
  effect of land surface roughness, 212
  effects of, by water, 224–5
  energy for water erosion, 210
  energy-intensity product of rainstorm, 211
  erosivity of rainstorm, 211
  factor of vegetation and land use, 213
  factors of, 208
  geological, accelerated, 206
  gully, 222
  gully areas of Nigeria, 22, 223–4
  human factor in, 213
  index of soil loss, 216–18
  management factor in, 215
  nature of, 207
  on steep slopes, 58
  part of hydrological-cum-rock cycle, 207
  rainfall erosion indices, 211
  relationship soil loss slope length, 212
  roads, footpaths, 216
  sheet, rill, 221
  soil detachability, 219
  soil structural stability, 208–9
  splash, 219, 225
  surface drainage characteristics affecting, 212

Soil fertility, 1
  chemical basis of, 192
  exchangeable bases, 203
  factors of, 192
  heat and air, 194–6
  mechanical support, 192–4
  physical basis of, 192
  problem of, 3
  properties, functions of soil organic matter, 198–9
  soil tilth, 192, 198
  water supply, 196
Soil formation
  active factors of, 50
  biosphere and, 54–5
  calcification, 47, 66, 70, 75, 138, 179
  climate, energy and, 50
  concept of periodicity and polygenesis, 63
  concept of steady state, 63
  correlation of processes and reactions with precipitation, 51–2
  eluviation, illuviation, 67
  ferralitization, 47, 66, 171, 178
  gleization, 66, 70, 76–7, 154
  humid tropical environment and, 157, 159–60
  humification, 54, 66, 68–9, 173, 185
  k-cycle, concept of, 138
  lateritization, 66, 70, 71–2, 154, 158, 164, 168
  leaching, 66
  mineralization, 69
  nitrogen fixation, cycling, 69–70
  parent material and, 56–7
  passive factors, 50, 55
  podzolization, 47, 66, 70, 73–5, 142, 154, 158, 173
  precipitation and, 50–3
  processes of, 65–6
  role of man, 55
  role of organisms, 55
  salinization, 47, 66, 70, 75–6, 179, 183
  simple processes, 66
  soil periodicity, 142
  soil–soil formers equation, 48–9
  soil water balance, 51
  solodization, 66, 70, 154, 179, 181, 183
  solonization, 179, 181, 183
  stages of, 33
  temperature and, 53–4
  time factor in, 61, 63–4
  topography and, 57–61
  Zakharov's system of classification, 70
Soil geography, 237
Soil map of Africa, 191
  Inter-African Pedological Service, 144
  legend of, 146–7
  problems of integration, 144–5
Soil microbiology, field of, 1
Soil moisture
  gypsum block method, 117
  methods of determination, 117
Soil monolith, continuous, discontinuous, 122

Soil orders
  azonal, 128, 129, 137, 139
  intrazonal, 128, 129, 137, 143
  zonal, 128, 129, 137, 143
Soil organic matter, 9, 12–14
  as store of plant nutrients, 199
  different types of, 13, 119–20
  humic acid, 198
  litter, aspects measured in field, 118–19
  organic nutrients, 10
  organic polymers, 27, 199
  plant roots, description, 119
  role of, 27
Soil peds, type and size description, 114–15
Soil pH
  determination by colorimetric method, 121
  determination by potentiometric method, 120
Soil physics, field of, 11
Soil–plant relations, 3
Soil porosity
  definition, 22
  factors influencing, 22
Soil profile
  characteristic layering, 29
  deep weathering, 32
  history of A, B, C, 32
  measurement of soil depth, 32
  unit of soil study, 97
Soil profile description
  aspects of horizon boundaries defined, 106–7
  classification, nomenclature of horizons, 106
  coarse fragments, abundance, 111–12
  coarse fragments, mineralogy, 114
  coarse fragments, shape, 114
  concentration of free carbonates, 121, 122
  faunal influence, 120
  observable secondary minerals, 121
  organic matter, 118
  porosity, 116
  preparation of profile pit, 105
  soil colour, 107
  soil moisture, drainage, 117
  soil pH determination in field, 120
  soil texture determination, 108
  stone sizes, 112, 113
  structure, constitution, 113
Soil profile, field study of
  cutting soil monolith, 98, 123
  field sheets, proformas, 99
  soil profile characteristics, 99, 105
  soil site characteristics, 99, 100
  soil survey equipment, 97–8
Soil properties, 17–30
Soil reaction
  concentration of hydroxyl anions, 28
  defined, 27

determined by, 28
pH range of soils, 28
pH scale and meter, 28
sources of hydrogen ions, 28
Soil science, 1, 3, 237
  Journal of, 6
  Russian period of, 4
  Society of America, proceedings, 6
Soil site characteristics
  climate, 105
  land use, 104
  location, 101
  measurement of slopes, 101–102
  parent material, 103
  site drainage, 103
  topographic data, 101
  vegetation, 104
Soil structure
  cementing substances, 21
  consistence when dry, 115
  consistence when moist, 116
  consistence when wet, 116
  defined, 21
  description of soil peds, 21
  factors of development, 21
  role of calcium, 21
  role of organic matter, 21
  role of sodium, 21
Soil survey, 6
  integrated land resources survey, 89
  manuals, 80, 97
  purpose, 78
Soil survey in Africa
  catena concept, 92
  CCTA soils map of Africa, 92
  INEAC classification system, 95
  IAR, Samaru, Zaria, 93, 94
  LRD in Africa, 92–3
  ORSTOM system, 95
  problems of, 91–3
  western Nigeria, 92, 93–4
Soil survey interpretation
  aim, 86
  FAO grouping of Nigerian soils, 87–9
  interpretative soil maps, 87
  land capability classification system, 87
Soil survey organizations
  central soil-testing laboratory, 80
  history of development, 79
  manual of soil survey, 80
  unit, 79–80
Soil survey project
  detailed survey, 83
  field survey, 81–2
  interpolation and extrapolation methods, 83
  office and laboratory work, 85–6
  pedological air photoanalysis, 82–3
  reconnaissance, 82
  semi-detailed surveys, 82
  soil survey memoir, 86

soil surveyor's approach in the field, 83
stages in, 80
types of field observations, 81
Soil temperature
  as regulating soil forming processes, 22
  factors of, 196
  freezing and heaving, 53–4
  frost and freeze-thaw action, 196
  influences, 196
  linked with soil air, 195
  specific heat of soils, 22–3
  thermal conductivity, 23
Soil texture
  defined, 18
  determination in the field, 19
  influence on soil properties, 18
  systems of texture grades, 18–19
  textural classes, diagnosis, 108–11
Soil type, 6
Soil water
  and available water, 197, 198
  and leaching, 52
  as carrier of substances, 51
  capillary rise, 151
  capillary translocation, 52
  capillary water, 15–16
  control of, by man, 198
  field capacity, defined, 15, 197
  free-draining water, 15, 198
  hygroscopic coefficient of soil, 16
  infiltration capacity, 15
  ionization of elements in solution, 196
  loss of, 16
  permanent wilting coefficient, 198
  permanent wilting percentage, 16
  plants' tolerance to water, 196–7
  singular characteristic of, 196
  soil solution, 15
  superfluous water, 197–8
  Thornthwaites's precipitation effectiveness index, 52–3
  types of, 15
  unavailable water, 198
  wilting coefficient of soil, 16, 198
  wilting point, 16

.oility to support crop production, 3
genetic classification of, 5
laboratory analysis of, 237–63
natural laws in respect of, 5
of cold lands, 184
of desert regions, 183
of humid temperate regions, 172
of humid tropics, 157
of interior grasslands, 181
of semi-arid and arid regions, 179
relict, 186
study and classification of, 4
zonality of, 5

Soils, semi-arid and arid, 179–84
  desert loam, 135, 184
  desert sand, 135, 184
  rendzina, 135, 175
  sodolic soils, 76, 135, 181, 183, 191
  solodized-solonetz, 76, 135, 181, 191
  solonchaks, 76, 135, 180, 183
  solonetz, 76, 135, 181, 183, 184
Soils of cold lands
  acid moor, 185, 186
  alkaline peats, 186
  alpine humus, 185, 186
  bog, 185
  half bog, 185
  meadow tundra, 185
  peat, 185
  peaty gley, 185, 186
  tundra gley, 185
Soils of interior grasslands
  black earth, 135, 181, 182
  chernozem, 135, 181, 182, 191
  chestnut, 181, 182–3
  prairie, 135, 181
  sierozem, 135, 181
Solifluction, 184, 185
Stace, H. C. T., 132
Stallings, J. H., 228
Stephens, C. G., 56, 78
Stewart, G. A., 91
Sulphate soils, 3
Sulphur
  cycle, 202
  mineral fertilizers with, 203
  production of amino-acids, 201
  sources in the soil, 202
  sulphur-demanding crops, 203
Sydney district, Australia, 56, 162, 163, 166
Sykes, R. A., 232
Symbiotic bacteria, 195, 200
Systematic classification of climate, 5

Taylor, J. A., 122
Terra rosa, 135, 179
Thiobacillus thiooxidans, 202
Thomas, M. F., 91
Thorp, J., 129
Titanium, 10
Titanium oxide, 161
Trace elements, 10
Tuley, P., 93

USDA, soil survey manual, 6, 80

Vanadium, 10
Van Baren, F. A., 160
Vilensicii, D. G., 5, 8
Vine, H., 93

Water, as proportion of living weight of plants, 196
Watts, D., 196
Weathering, processes of, 1
Whitney, M., 6, 79
Whitt, D. M., 211

Wind erosion, 212
  damage from, 228
  factors which affect, 227
  forms of particle movement, 226–7
  process of, 1, 225
  role of wind barriers, 227
  sand dunes and loess, 227
  threshold wind velocity, 225–6
Wischmeier, W. H., 216
Wright, R. L., 72, 165

Yaalon, D. H., 36
Yair, A., 36

Zakharov, 70
Zinc, 10
Zingg, A. W., 211, 212
Zircon, 46
Zonal concept, 5